Springer Theses

Recognizing Outstanding Ph.D. Research

Aims and Scope

The series "Springer Theses" brings together a selection of the very best Ph.D. theses from around the world and across the physical sciences. Nominated and endorsed by two recognized specialists, each published volume has been selected for its scientific excellence and the high impact of its contents for the pertinent field of research. For greater accessibility to non-specialists, the published versions include an extended introduction, as well as a foreword by the student's supervisor explaining the special relevance of the work for the field. As a whole, the series will provide a valuable resource both for newcomers to the research fields described, and for other scientists seeking detailed background information on special questions. Finally, it provides an accredited documentation of the valuable contributions made by today's younger generation of scientists.

Theses are accepted into the series by invited nomination only and must fulfill all of the following criteria

- They must be written in good English.
- The topic should fall within the confines of Chemistry, Physics, Earth Sciences, Engineering and related interdisciplinary fields such as Materials, Nanoscience, Chemical Engineering, Complex Systems and Biophysics.
- The work reported in the thesis must represent a significant scientific advance.
- If the thesis includes previously published material, permission to reproduce this must be gained from the respective copyright holder.
- They must have been examined and passed during the 12 months prior to nomination.
- Each thesis should include a foreword by the supervisor outlining the significance of its content.
- The theses should have a clearly defined structure including an introduction accessible to scientists not expert in that particular field.

More information about this series at http://www.springer.com/series/8790

Rui Shang

New Carbon–Carbon Coupling Reactions Based on Decarboxylation and Iron-Catalyzed C–H Activation

Doctoral Thesis accepted by
the University of Science and Technology of China,
Hefei, China

 Springer

Author
Dr. Rui Shang
Department of Chemistry
University of Science and Technology
  of China
Hefei, Anhui
China

Supervisor
Prof. Yao Fu
Department of Chemistry
University of Science and Technology
  of China
Hefei, Anhui
China

ISSN 2190-5053   ISSN 2190-5061   (electronic)
Springer Theses
ISBN 978-981-10-9813-0   ISBN 978-981-10-3193-9   (eBook)
DOI 10.1007/978-981-10-3193-9

This Springer imprint is published by Springer Nature
The registered company is Springer Nature Singapore Pte Ltd.
The registered company address is: 152 Beach Road, #22-06/08 Gateway East, Singapore 189721, Singapore

I dedicate this thesis to my parents,
Jun Shang, and Sufen Qi, who offered me
unconditional love and support that have
always been there. Thank you so much!

Supervisor's Foreword

I am very pleased to see the doctoral thesis of my former student Dr. Rui Shang being published as a part of the Springer Theses. The thesis is very fruitful, containing a lot of achievements Rui made during his Ph.D. study. It includes the first part conducted at the University of Science and Technology of China about transition metal-catalyzed decarboxylative cross-couplings and the second part conducted in Japan under the supervision of Prof. Eiichi Nakamura, describing the development of catalytic uses of iron for C–H functionalization.

The development of new methodologies and new catalysts for catalytic carbon–carbon bond formation reactions is the central goal in modern synthetic organic chemistry. Despite the established methodologies that significantly improved the capability of human beings to create complex chemicals and also to solve environmental and energy-related problems, new versions of C–C coupling reaction utilizing easily available, environmentally friendly reagents and abundant, sustainable catalysts are still highly required. The first part of this thesis focuses on the developments of new catalytic C–C bond formation methods using easily accessible carboxylate salts through catalytic decarboxylation with good atom economy. Traditional cross-coupling methodologies largely rely on the use of organometallic reagents. Usually, the organometallic reagents are expensive and sensitive to air and moisture, which makes them difficult to prepare and store on a large scale, and they generate metal wastes as the side products. Although the use of carboxylic acid as surrogates through extrusion of carbon dioxide to generate carbon nucleophiles in cross-coupling has been reported by Nilson in 1960s, it was buried in the literatures without further development until 2000s. Myers and co-workers reported decarboxylative Heck reaction, and Goossen and co-workers reported Cu/Pd co-catalyzed biaryl synthesis, which reinvigorated interest in this area. Until recently, this research area is still expanding, and many cost-effective reactions are being developed. This part of Rui's Ph.D. work contributed to the development in the versatility of novel and practical decarboxylative cross-couplings. He ingeniously discovered a series of decarboxylative couplings using easily accessible and cheap carboxylate salts such as oxalate salts, cyanoacetate salts, and perfluorobenzoates to synthesize high value-added chemicals. Mechanistic studies to understand the elementary step of metal-catalyzed decarboxylation were also carried out through the collaboration with theoretical chemists. The second part of this thesis was conducted in Prof. Eiichi Nakamura's laboratory at the University of Tokyo. I was pleased to see Rui performed well under the guidance of Prof. Nakamura. Iron is the most suitable element conforms to the pirit of the "Elements Strategy Initiative" for catalysis due to its ubiquity and non-toxicity. This part of research is devoted to exploring the catalytic reactivity of organoiron species and hence to developing synthetically useful carbon–carbon formation reactions through direct activation of C–H bonds. The key achievements of this part of research are the discoveries of several methods to tame the catalytic reactivity of organoiron, which enable various types of C–H transformations. The detailed achievements include the following: (1) discovery of bisphosphine ligand in combination with a bidentate auxiliary, which enables the first example of iron-catalyzed $C(sp^3)$-H functionalization through ferracycle intermediate and (2) discovery of a B-Zn transmetalation reaction to enable the direct use of organoboronate reagents in iron-catalyzed C–H functionalization to achieve versatility and efficiency rivaling precious metal catalysis. These discoveries and ensuing research activities brought Rui many honours, such as a Research Fellowship from the Japan Society for the Promotion of Science (2014); Student Presentation Award of the 94th CSJ Annual Meeting; Presidential Award of Chinese Academy of Sciences (2014); and Hundred Excellent Doctoral Thesis of the Chinese Academy of Sciences (2015). He was awarded the 7th National Award for Youth in Science and Technology by the Chinese government in the Great Hall of the People in Beijing. He is currently a researcher under the JSPS fellowship at the Department of Chemistry, the University of Tokyo, with Prof. Eiichi Nakamura.

On behalf of the Department of Chemistry at the University of Science and Technology of China, I would like to congratulate him on the Springer Theses Award, and I believe he will continue exploring new facets of chemistry and become a successful scientist in the future.

Hefei, China   Prof. Yao Fu
September 2016

Parts of this thesis have been published in the following journal articles:

1. "Synthesis of Aromatic Esters via Pd-Catalyzed Decarboxylative Coupling of Potassium Oxalate Monoesters with Aryl Bromides and Chlorides"

Shang, R.; Fu, Y.; Li, J. B.; Zhang, S. L.; Guo, Q. X.; Liu, L. *J. Am. Chem. Soc.* **2009**, *131*, 5738.

2. "Copper-Catalyzed Decarboxylative Cross Coupling of Potassium Polyfluorobenzoates with Aryl Iodides and Bromides"

Shang, R.; Fu, Y.; Wang, Y.; Xu, Q.; Yu, H.-Z.; Liu, L. *Angew. Chem. Int. Ed.* **2009**, *48*, 9350.

3. "Pd-Catalyzed Decarboxylative Cross Coupling of Potassium Polyfluorobenzoates with Aryl bromides, Chlorides, and Triflates"

Shang, R.; Xu, Q.; Jiang, Y.-Y.; Wang, Y.; Liu, L. *Org. Lett.* **2010**, *12*, 1000.

4. "Theoretical Analysis of Factors Controlling Pd-Catalyzed Decarboxylative Coupling of Carboxylic Acids with Olefins"

Zhang, S.-L.; Fu, Y.; **Shang, R.**; Guo, Q.-X.; Liu, L. *J. Am. Chem. Soc.* **2010**, *132*, 638.

5. "Palladium-Catalyzed Decarboxylative Couplings of 2-(2-Azaaryl) Acetates with Aryl Halides and Triflates"

Shang, R.; Yang, Z. W.; Wang, Y.; Zhang, S. -L.; Liu, L. *J. Am. Chem. Soc.*, **2010**, *132*, 14391.

6. "Synthesis of α Aryl Nitriles through Palladium Catalyzed Decarboxylative Coupling of Cyanoacetate Salts with Aryl Halides and Triflates"

Shang, R.; Ji, D. S.; Chu, L.; Fu, Y.; Liu, L.* *Angew.Chem. Int. Ed.* **2011**, *50*, 4470.

7. "Palladium-Catalyzed Decarboxylative Coupling of Potassium Nitrophenyl Acetates with Aryl Halides"

Shang, R.; Huang, Z.; Chu, L.; Fu, Y.; Liu, L. *Org. Lett.* **2011**, *13*, 4240.

8. "Transition Metal-Catalyzed Decarboxylative Cross-Coupling Reactions"

Shang, R.; Liu, L. *Science China (Chemistry).* **2011**, *54*, 1670.

9. "β Aryl Nitrile Construction via Palladium Catalyzed Decarboxylative Benzylation of α Cyano Aliphatic Carboxylate Salts"

Shang, R.; Huang, Z.; Xiao, X.; Lu, X.; Fu, Y.; Liu, L. *Adv. Syn. Catal.* **2012**, *354*, 2465.

x   Parts of this thesis have been published in the following journal articles:

10. "β-Arylation of Carboxamides via Iron-catalyzed C (sp^3)–H Bond Activation"

Shang, R.; Ilies, L.; Matsumoto, A.; Nakamura, E. *J. Am. Chem. Soc.* **2013**, *135*, 6030.

11. "Copper-Catalyzed Decarboxylative Coupling of Alkynyl Carboxylates with 1, 1-Dibromo-1-alkenes"

Huang, Z.; **Shang, R.;** Zhang, Z.-R.; Tan, X.-D.; Xiao, X.; Fu, Y. *J. Org. Chem.* **2013**, *78*, 4551.

12. "Iron-Catalyzed C (sp^2)–H Bond Functionalization with Organoboron Compounds"

Shang, R.; Ilies, L.; Sobi, A.; Nakamura, E. *J. Am. Chem. Soc.* **2014**, *136,* 14349.

Acknowledgements

First I would like to express my deepest appreciation to my family for nursing me with affection and unconditional love.

I would like to express my special appreciation and thanks to my advisor Prof. Yao Fu, Prof. Qingxiang Guo from the University of Science and Technology of China, and also Prof. Lei Liu from Tsinghua University, who have instructed me since I was a junior undergraduate. The first part of my research in this thesis was conducted under their supervisions. I would also like to thank all the members of the research group in USTC, especially Mr. Zheng Huang, Mr. Ling Chu, Mr. Dongsheng Ji, Dr. Jiabin Li, Mr. Yan Wang, Dr. Zhiwei Yang, Dr. Xi Lu, Dr. Yuanye Jiang, Prof. Song-Lin Zhang, and Prof. Haizhu Yu, who supported me in my research and made contributions to the works presented in this thesis.

I would like to thank the Department of Chemistry at the University of Science and Technology of China. I deeply appreciate all the teachers who taught me during my undergraduate and graduate studies, in particular Prof. Zude Zhang and Prof. Baozhong Zhang who taught me inorganic chemistry and organic chemistry for National Chemistry Olympics since I was in high school.

I express my sincerest gratitude to Prof. Eiichi Nakamura, who supervised my Ph.D. study from October 2012 to April 2014 at the Department of Chemistry, Graduate School of Science, the University of Tokyo, for his thorough professional guidance and valuable suggestions. I am also grateful to Associate Prof. Laurean Ilies for his daily advice and support. I am grateful to the Secretary Akemi Maruyama, who kindly helped me to make my stay in Tokyo more comfortable and convenient. During my exploration of iron-catalyzed C–H bond activation chemistry, I closely worked with Dr. Sobi Asako, Dr. Tatsuaki Matsubara, Dr. Masaki Sekine, Mr. Yuki Itabashi, Mr. Takumi Yoshida, Mr. Hiroki Sato, Ms. Saki Ichikawa, and Ms. Mayuko Isomura. I would like to thank all of them for their help in both research and daily life.

Mr. Ming-Chen Fu and Mr. Wan-Min Cheng from USTC are acknowledged for their help in translating and editing this thesis for the publication in Springer Theses.

Finally, when I have summarized all my works done in the past 6 years in this thesis, I am feeling quite fulfilled. The completion of this thesis is a new start in my life. I will continue to make new discoveries and explore the unknown world of chemistry in the future.

Contents

Part I
New Carbon–Carbon Coupling Reactions
Based on Decarboxylation

Chapter 1
Transition Metal-Catalyzed Decarboxylation and Decarboxylative Cross-Couplings

Abstract Transition metal-catalyzed decarboxylative cross-coupling reactions have recently emerged as a new and important category of organic transformations that find versatile applications in the construction of carbon–carbon and carbon-heteroatom bonds. The use of relatively cheap and stable carboxylic acids to replace organometallic reagents enables the decarboxylative cross-coupling reactions to proceed with good selectivities and functional group tolerance. In the present review, we summarize the various types of decarboxylative cross-coupling reactions catalyzed by different transition metal complexes. The scope and applications of these reactions are described. The challenges and opportunities in the field are discussed.

1.1 Introduction

Transition metal-catalyzed cross-coupling reactions have fundamentally changed the strategies in organic synthesis. They have greatly enhanced our ability to synthesize complex organic molecules and found extensive applications in the preparation of natural products and functional materials [1]. Nonetheless, in most of the transition metal-catalyzed cross-coupling reactions there is a need to use prefunctionalized organometallic reagents, which are usually expensive, highly toxic, and unstable. This creates an important challenge to develop more stable and less expensive reagents that are applicable to transition metal-catalyzed cross-coupling reactions.

As a reaction appeared in basic organic chemistry, decarboxylation reaction has been extensively studied in the past. However, metal-catalyzed decarboxylation of carboxylic acid compounds to produce metal–carbon bonds as d-synthon equivalent, which subsequently participates in the new bonds forming process, has not received enough attention until the inspiring works of Myers [2] and Gooßen [3] in recent years. Such reactions use stable carboxylic acids (or their salts) as substrates and generate organometallic intermediates through the removal of CO_2. An important advantage of the decarboxylative cross-coupling reactions is the avoidance of using any strong base for transmetalation. Thus these reactions often show good tolerance to many functional groups. After intensive studies in recent years,

© Springer Nature Singapore Pte Ltd. 2017
R. Shang, *New Carbon–Carbon Coupling Reactions Based on Decarboxylation and Iron-Catalyzed C–H Activation*, Springer Theses,
DOI 10.1007/978-981-10-3193-9_1

decarboxylative cross-coupling reactions have become an important category of transformations in organic synthesis. In this part, we briefly survey the scope of this type of reaction. Note that we focus on the decarboxylative cross-coupling reactions that proceed through the formation of carbon–metal bonds.

1.2 Metal-Catalyzed Decarboxylation Reactions

Decarboxylation catalyzed by transition metal complex to form C–M bond is the critical step in decarboxylative cross-coupling reactions. At present, several metals have been proved to be able to catalyze redox-neutral decarboxylation, such as Cu, Ag, Pd, Au, and Rh.

1.2.1 Cu-Catalyzed Protodecarboxylation

Copper complexes have the ability to promote decarboxylation of carboxylic acids. Since the 1960s, Nilsson [4], Cohen [5], Shepard [6], and others reported that some Cu complexes promoted the protodecarboxylation of aromatic carboxylic acids (Scheme 1.1).

More recent studies by Goossen et al. confirmed the earlier findings [7], showing that Cu(I) complexes with phenanthroline as supporting ligand can catalyze the protodecarboxylation of many aromatic acids under relatively mild conditions (Scheme 1.2). Moreover, the use of microwave heating can further accelerate the protodecarboxylation process [8].

Another type of acid that can undergo protodecarboxylation in the presence of copper catalyst is alkynyl carboxylic acid. As shown in Scheme 1.3, these acids can be converted to alkynes under fairly mild conditions [9].

1.2.2 Ag-Catalyzed Protodecarboxylation

The Goossen group [10] and Larrosa group [11] independently reported the Ag(I)-catalyzed protodecarboxylation of aromatic acids and heteroaryl acids. These

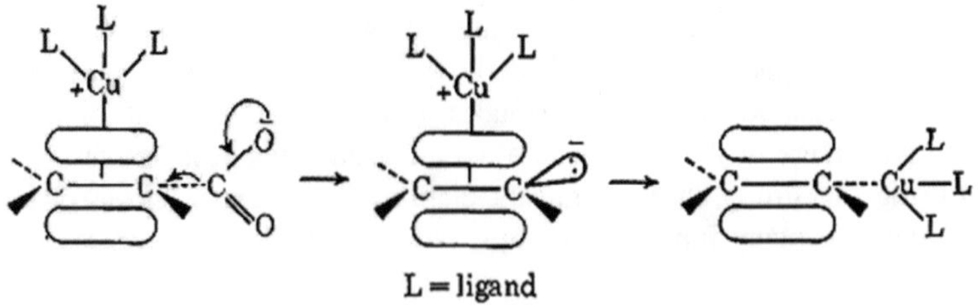

Scheme 1.1 Copper complex promoted decarboxylation

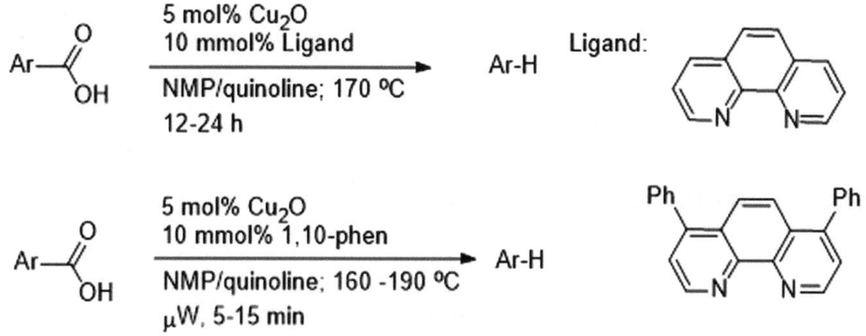

Scheme 1.2 Copper/Phenanthroline complex catalyzed protodecarboxylation of Ar–COOH

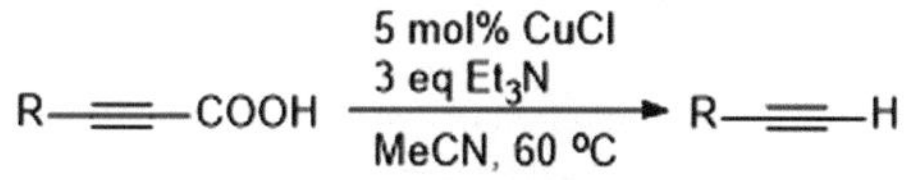

Scheme 1.3 Copper-catalyzed protodecarboxylation of alkynyl carboxylic acids

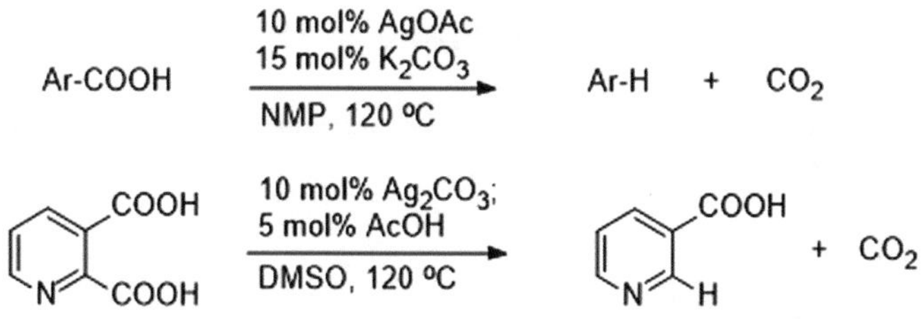

Scheme 1.4 Silver-catalyzed protodecarboxylation of aromatic carboxylic acids

reactions proceed under milder conditions as compared to their Cu(I)-catalyzed counterparts, indicating that the Ag(I) salts are more effective in activating the decarboxylation of some carboxylic acids (Scheme 1.4). This conclusion is consistent with the results of a theoretical study carried out by Gooßen et al. [12].

1.2.3   Au-Catalyzed Protodecarboxylation

Au is in the same group with Cu and Ag. However, Au complexes have been regarded as inert in the past several decades as catalyst for organic synthesis and few studies have been carried out to examine whether gold complexes can catalyze decarboxylation due to its high cost. Very recently Nolan et al. reported the first example of the Au(I)-mediated decarboxylation of aromatic acids [13]. Surprisingly, the products are not protonated because the reaction stops at the metalation (i.e., auration) stage (Scheme 1.5). This observation may be explained by the high stability of the carbon–Au bond.

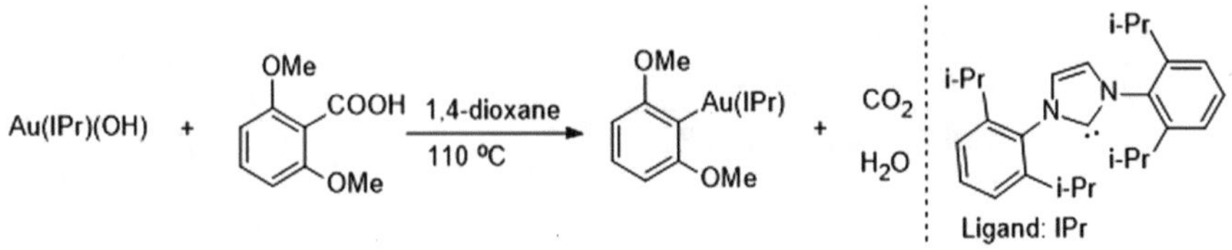

Scheme 1.5 Gold(I) mediated decarboxylative auration reported by Nolan et al.

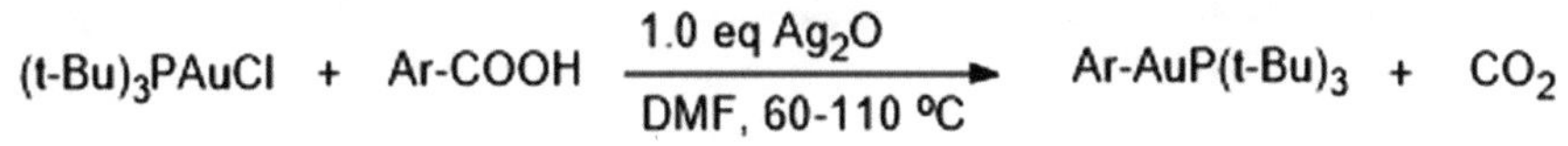

Scheme 1.6 Gold(I) mediated decarboxylative auration reported by Larrosa et al.

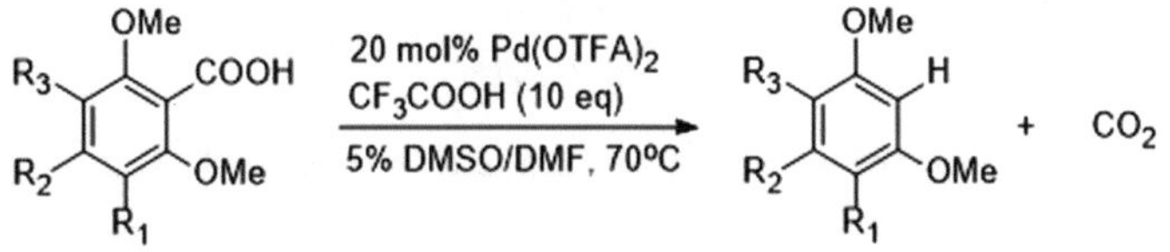

Scheme 1.7 Palladium-catalyzed protodecarboxylation reported by Kozlowiski et al.

Larrosa et al. [14] also reported similar decarboxylative auration reactions mediated by $(t\text{-Bu})_3\text{PAuCl}$ (Scheme 1.6).

1.2.4 Pd-Catalyzed Protodecarboxylation

Decarboxylation catalyzed by palladium complexes also has been reported. Kozlowski et al. reported Pd(II)-catalyzed protodecarboxylation of aromatic carboxylic acid under 70 °C (Scheme 1.7) [15]. Unfortunately, this reaction requires the use of a fairly large amount of Pd catalyst, and the substrate scope of the reaction is also very limited.

A related study by Myers and coworkers showed the formation of the Pd(II) carboxylate complex with the carboxylate sodium salt. This complex undergoes decarboxylative metalation to generate an aryl-Pd(II) intermediate as shown in Scheme 1.8 [16].

1.2.5 Rh-Catalyzed Protodecarboxylation

Recently, Zhao et al. reported the Rh (I)-catalyzed decarboxylative protonation reactions of activated carboxylic acids to cleave $C(sp^2)$–COOH or $C(sp^3)$–COOH bonds in nonpolar toluene solvent under relatively mild conditions (Scheme 1.9) [17].

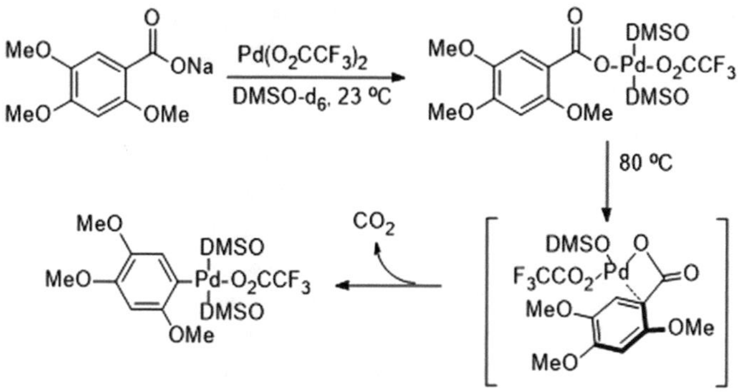

Scheme 1.8 Decarboxylative palladation reported by Myers and coworkers

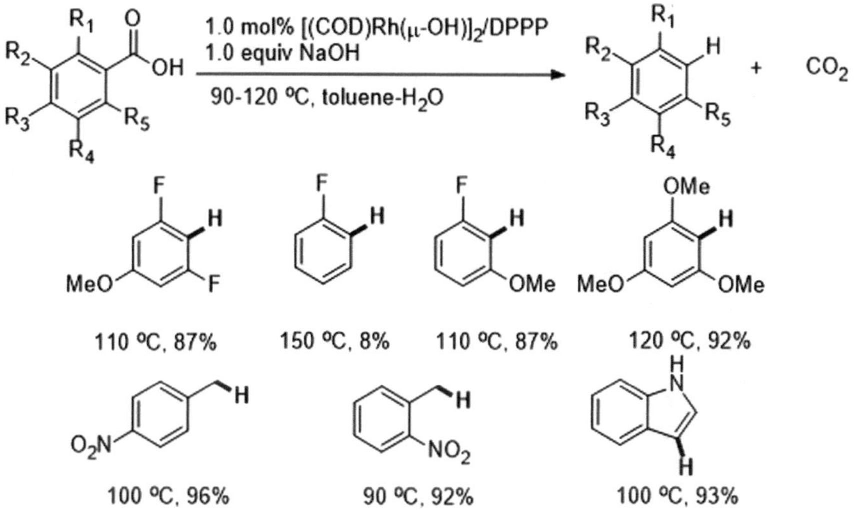

110 °C, 87%   150 °C, 8%   110 °C, 87%   120 °C, 92%

100 °C, 96%   90 °C, 92%   100 °C, 93%

Scheme 1.9 Rh-catalyzed protodecarboxylation reported by Zhao and coworkers

1.3 Decarboxylative C–C Cross-Coupling Reactions

1.3.1 Cross-Coupling Between C–COOH and C–X Bonds

1.3.1.1 Biaryl Formation (Suzuki and Stille Type Decarboxylative Cross-Coupling Reactions)

Biaryls are important structural motifs in many biologically active compounds. Reports have shown that biaryls can be obtained through decarboxylative cross-coupling of aromatic carboxylic acids with aryl halide [18]. A number of catalyst systems have been developed to accomplish the reaction. These reactions include:

1. Pd/Cu bimetallic systems

In 2006 Gooßen et al. reported Pd/Cu-catalyzed decarboxylative cross-couplings of ortho-substituted aromatic carboxylic acids with aryl halides [3]. Two catalyst systems were examined. The first system contains a catalytic amount of Pd with a stoichiometric amount of Cu. This system can catalyze decarboxylative cross-coupling reaction of a series of 2-substituted aryl and alkenyl carboxylic acids with aryl halides (chloro, bromo) at 120 °C. The second catalytic system is composed of catalytic amount of palladium salt and copper salt, but the reaction temperature is higher (160 °C) due to the reduced amount of copper (Scheme 1.10).

In the subsequent studies, Goossen et al. optimized the catalysts and ligands, and extended the above reaction to aryl chlorides [19], triflates [20], and tosylates [21] (Scheme 1.11). It was found that the chloride or bromide anion generated in the reaction hampers the Cu-catalyzed decarboxylation process [22]. Therefore, when aryl triflates are used, the substrate scope can be extended to some less reactive aromatic carboxylates carrying no *ortho*-substituent.

As to the mechanism of the bimetallic catalytic system, Goossen et al. proposed that Cu plays a key role in the decarboxylation step (Scheme 1.11). An aryl-Cu intermediate is generated through decarboxylation, which participates in the Pd-catalyzed cross-coupling with aryl halide to generate new C–C bonds through transmetalation and reductive elimination. A Pd(0) complex is regenerated after the reaction, which can reenter the catalytic cycle through oxidative addition with aryl halides (Scheme 1.12).

2. Pd/Ag bimetallic systems

Becht et al. reported the Pd(II)-catalyzed decarboxylative cross-coupling of substituted aromatic carboxylates with aryl iodides [23] or diaryliodonium salts [24]. This reaction requires the use of a stoichiometric amount of Ag_2CO_3 (Schemes 1.13 and 1.14).

A similar reaction was also reported in 2009 by Wu et al. (Scheme 1.15) [25].

In 2010 Greaney et al. reported the decarboxylative cross-coupling reaction of substituted oxazole- or thiazole-5-carboxylic acids with aryl bromides and iodides.

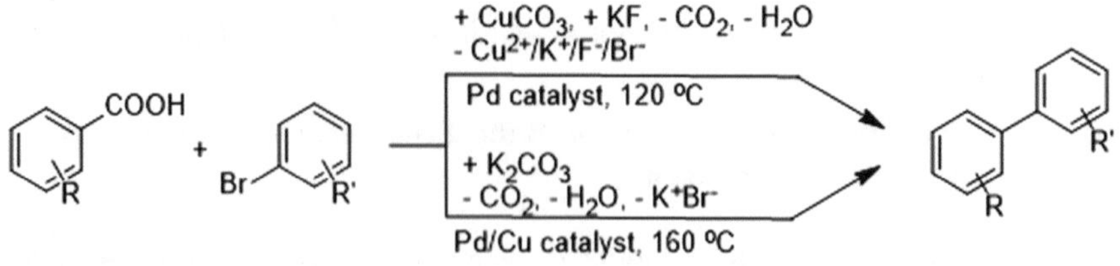

Scheme 1.10 Cu–Pd decarboxylative biaryl synthesis reported by Goossen et al.

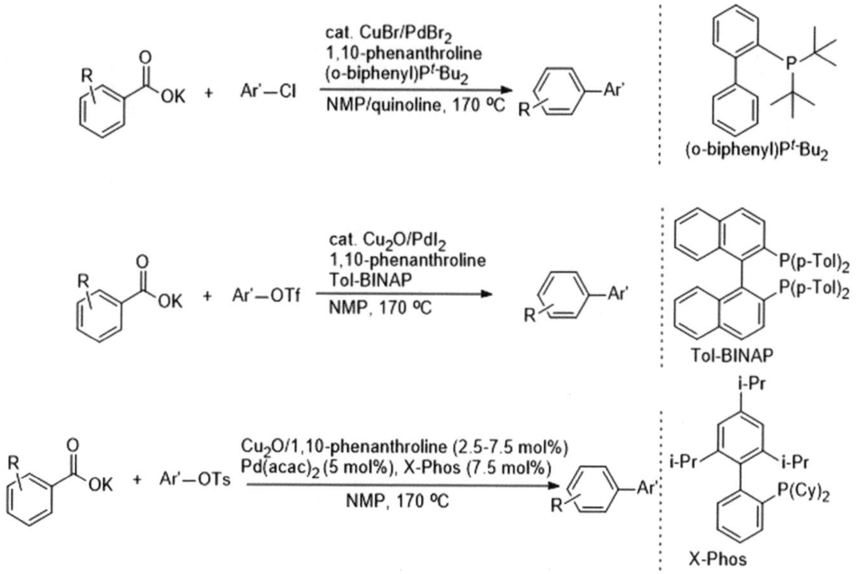

Scheme 1.11 Cu–Pd decarboxylative biaryl synthesis using Ar–Cl, Ar–OTf, Ar–OTs as electrophile

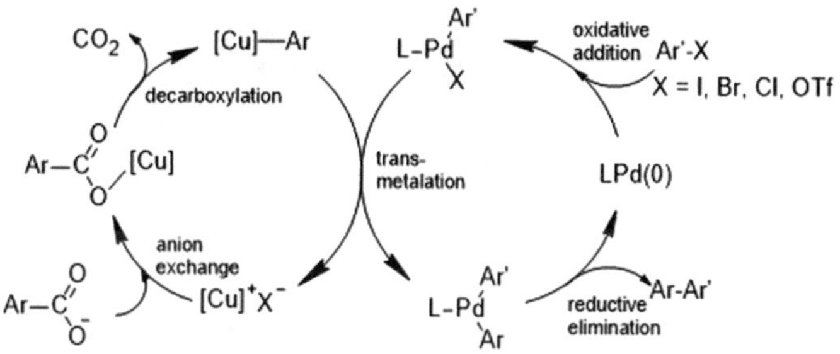

Scheme 1.12 Proposed mechanism for Cu–Pd-catalyzed decaboxylative biaryl synthesis

Again a Pd/Ag bimetallic catalyst system was used. This reaction produces 5-arylated oxazoles and thiazoles (Scheme 1.16) [26].

The above Pd/Ag bimetallic catalyst systems require a stoichiometric amount of Ag salt. This may be explained by the formation of insoluble and therefore, catalytically inactive silver halides in the reaction. To overcome this problem, Goossen et al. used aryl triflates as the electrophiles and successfully developed the first examples of the decarboxylative cross-coupling reactions catalytic in both Pd and Ag [27]. Because for some substrates, silver complexes possess a higher decarboxylating activity than copper complexes, the Pd/Ag-catalyzed decarboxylative

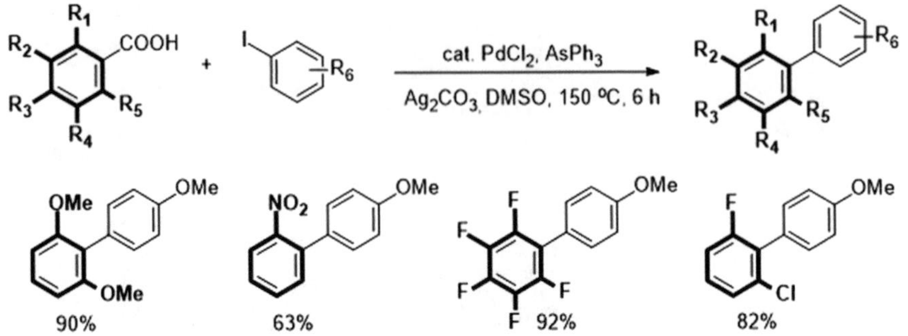

Scheme 1.13 Pd-catalyzed, silver-promoted decarboxylative biaryl coupling reported by Becht and Wagner et al.

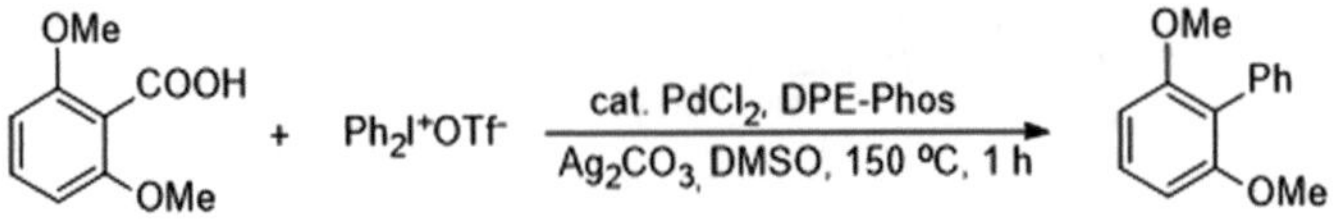

Scheme 1.14 Pd-catalyzed, silver-promoted decarboxylative biaryl coupling using diaryliodonium triflates

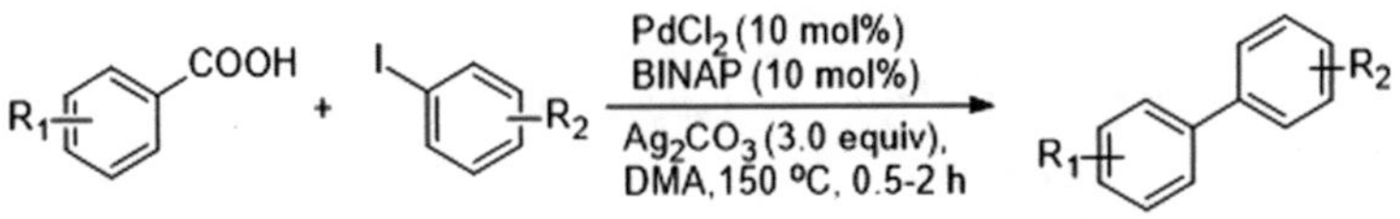

Scheme 1.15 Pd-catalyzed, silver-promoted decarboxylative biaryl coupling reported by Wu et al.

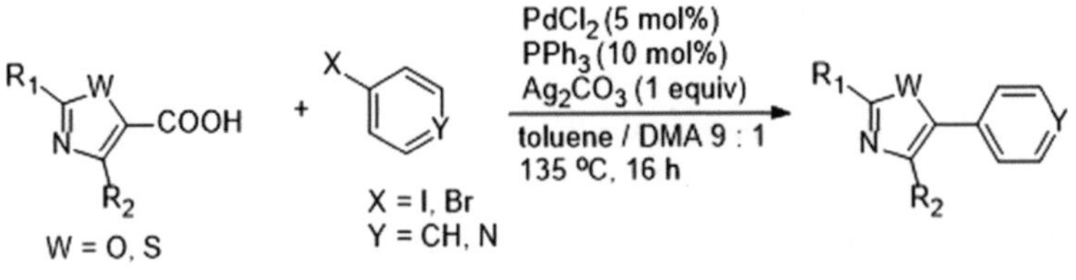

Scheme 1.16 Decarboxylative coupling of azoyl carboxylic acids with aryl halides

cross-coupling reactions can proceed under milder conditions as compared to the Pd/Cu-catalyzed ones (Scheme 1.17).

3. Pd-only systems

Since many Pd complexes are good catalysts for the decarboxylative reactions, and also have high ability to catalyze the bond formation. As a result, using palladium

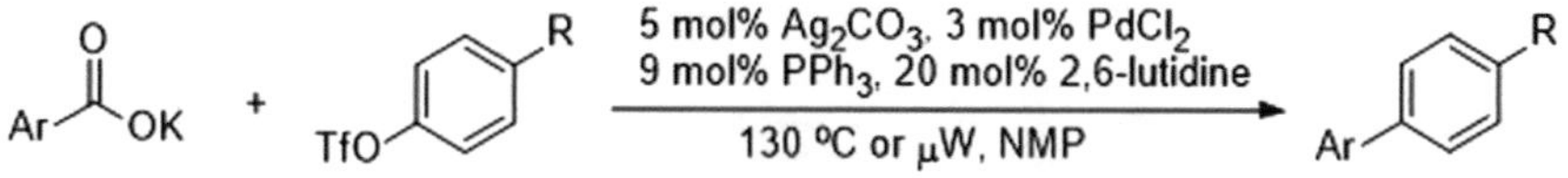

Scheme 1.17 Ag/Pd-catalyzed decarboxylative biaryl coupling reported by Goossen et al.

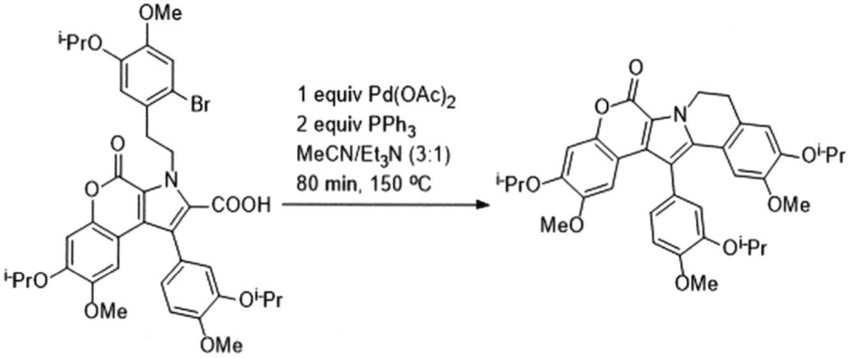

Scheme 1.18 Palladium mediated intermolecular biaryl synthesis reported by Steglich et al.

complex only can accomplish some decarboxylative cross-coupling reactions. Steglich and coworkers first reported the intramolecular decarboxylative cross-coupling in their total synthesis of lamellarin L using an equivalent amount of palladium salt (Scheme 1.18) [28].

Forgione and Bilodeau et al. studied the reactions of a number of heteroaromatic carboxylic acids (furoic carboxylic acid, thiophene carboxylic acid, and N-methyl pyrrole carboxylic acid) with aryl halides (Scheme 1.19) [29]. The mechanism of the reaction was proposed to involve the following steps. First, the Pd(0) catalyst reacts with aryl halides via oxidative addition to generate an aryl-Pd(II) intermediate. Then this Pd(II) intermediate forms a coordination complex with the carboxylate. Through decarboxylative palladation a bisarylated Pd(II) complex is produced. The final step is reductive elimination to form the desired C–C bond and regenerate the Pd(0) catalyst (Scheme 1.20).

In Forgione and Bilodeau's study, they found that the decarboxylative cross-coupling reactions show selectivity over C–H activation. For instance, when 3-methylthiophene is treated with an aryl halide under Pd catalysis, both 2- and 5-arylated products are obtained. In comparison, when 3-methylthiophene-2-carboxylic acid is treated with an aryl halide, the arylation only occurs at the 2-position, when 3-methylthiophene-5-carboxylic acid is tested, arylation at the 5-position is observed (Scheme 1.21).

Liu et al. found that polyfluorinated benzoic acids can undergo Pd-catalyzed decarboxylative cross-coupling with aryl bromides, chlorides, and even triflates [30]. It is critical to the success of the reaction that *ortho* position should be replaced by fluorine group or other electron-withdrawing groups (Cl^-, CF_3^-).

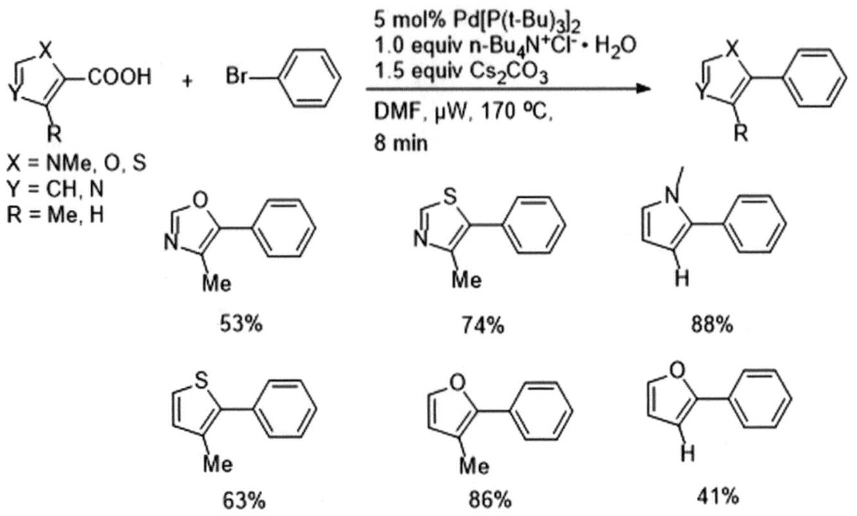

X = NMe, O, S
Y = CH, N
R = Me, H

53%   74%   88%

63%   86%   41%

Scheme 1.19 Pd-catalyzed decarboxylative arylation of heteroaromatic carboxylic acids with PhBr

Scheme 1.20 Mechanism postulated by Forgione and Bilodeau et al.

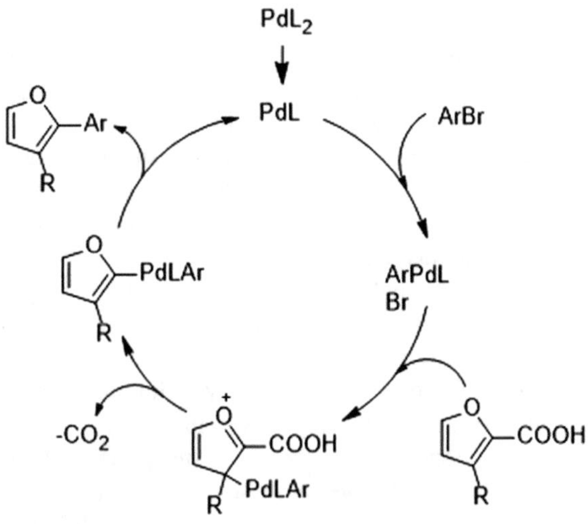

The reaction first undergoes an oxidative addition step, then carboxylate anion exchanges, and decarboxylation occurs in the divalent palladium complexes containing a phosphine ligand, and thus the reductive elimination process will take place. Through theoretical calculations, the author found that decarboxylative palliations constitute the rate-limiting step of the catalytic cycle (Scheme 1.22).

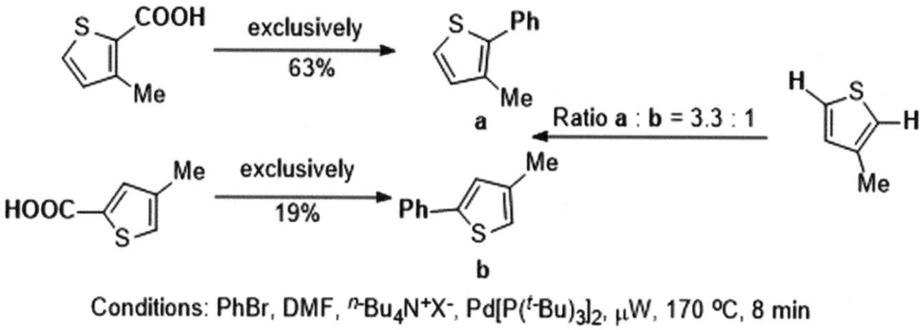

Conditions: PhBr, DMF, ^{n}-Bu$_4$N$^+$X$^-$, Pd[P(^{t}Bu)$_3$]$_2$, μW, 170 °C, 8 min

Scheme 1.21 Regioselectivity in decarbocylative arylation of heteroaromatic carboxylic acids

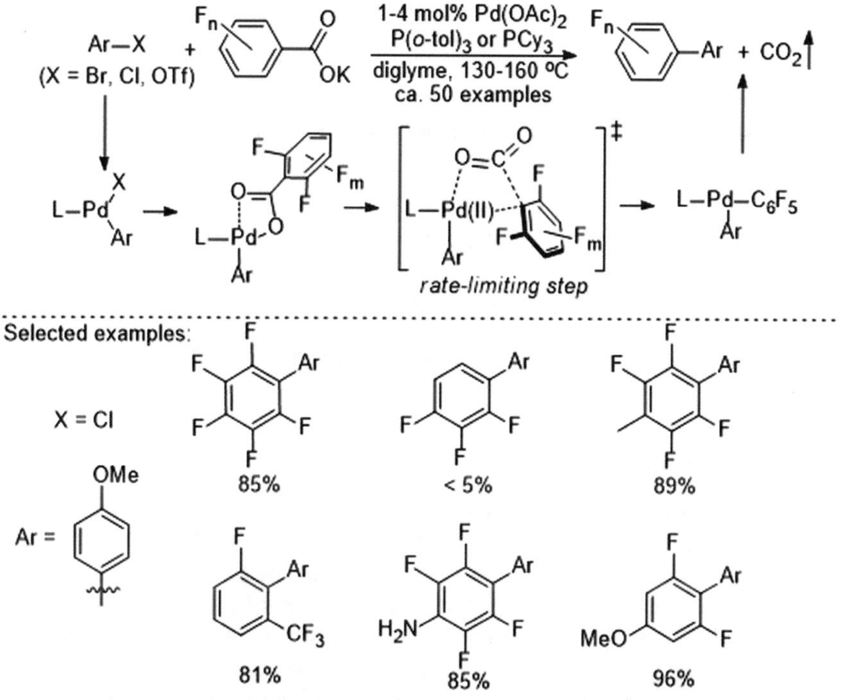

Scheme 1.22 Decarboxylative arylation of potassium perfluorobenzoates reported by Liu et al.

Miura et al. reported that 2-aryl-indole-3-carboxylic acids underwent Pd-catalyzed decarboxylative cross-coupling with aryl bromides (Scheme 1.23) [31].

Reynolds et al. showed that Pd can also catalyze the decarboxylative cross-coupling of pyrrole-2-carboxylic acids with aryl bromides (Scheme 1.24) [32].

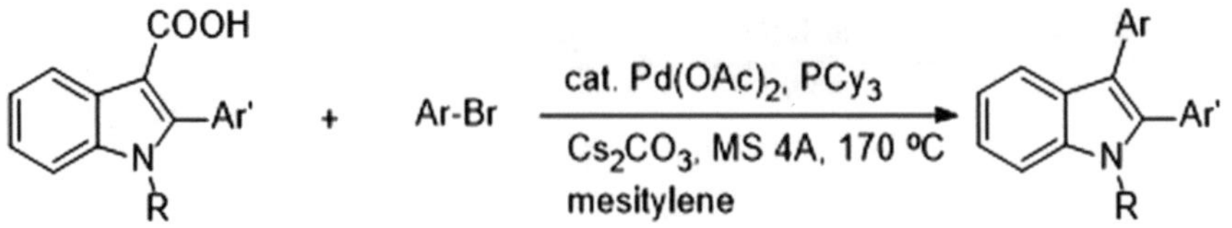

Scheme 1.23 Decarboxylative arylation of indole-3-carboxylic acid

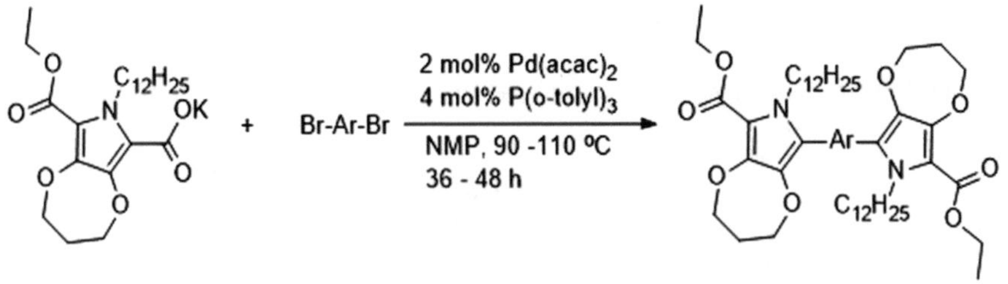

Scheme 1.24 Decarboxylative cross-coupling involving a 3,4-dioxypyrrole reported by Reynolds et al.

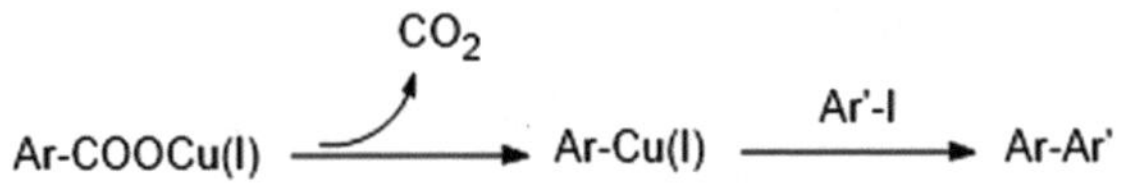

Scheme 1.25 Pioneering studies of Cu-mediated decarboxylative biaryl couplings by Nilsson et al.

4. Cu-only systems

In the 1960s Nilsson et al. found that Cu salts can promote decarboxylation of aromatic carboxylic acids to produce the aryl-Cu complexes, which are able to react with an aryl iodide to produce a biaryl [4] (Scheme 1.25).

This early example is stoichiometric in the presence of copper catalyst, whereas the first catalytic example for the use of Cu in decarboxylative cross-coupling was reported by Shang and Liu et al. in 2009. This is an efficient way to prepare the asymmetric polyfluorinated biaryls from polyfluorinated benzoic acids and aryl halides (aryl bromides and iodides) [33]. But to fluorinated acid, only the 2,6-difluoro acid or the 2,6-bis(trifluoromethyl) acid can get a satisfying yield (Scheme 1.26a). Through theoretical analysis, Liu et al. proposed that decarboxylation occurs with the Cu(I) complex of the polyfluorinated aromatic acid, generating an aryl-Cu(I) intermediate. This aryl-Cu(I) intermediate then reacts with the aryl halide through oxidative addition to form a bisarylated Cu(III) complex. Finally, reductive elimination takes place to produce the biaryl and regenerate the Cu(I) catalyst. It should be noted that the mechanism of Cu-catalyzed reactions and Pd-catalyzed reactions are totally different (Scheme 1.26b).

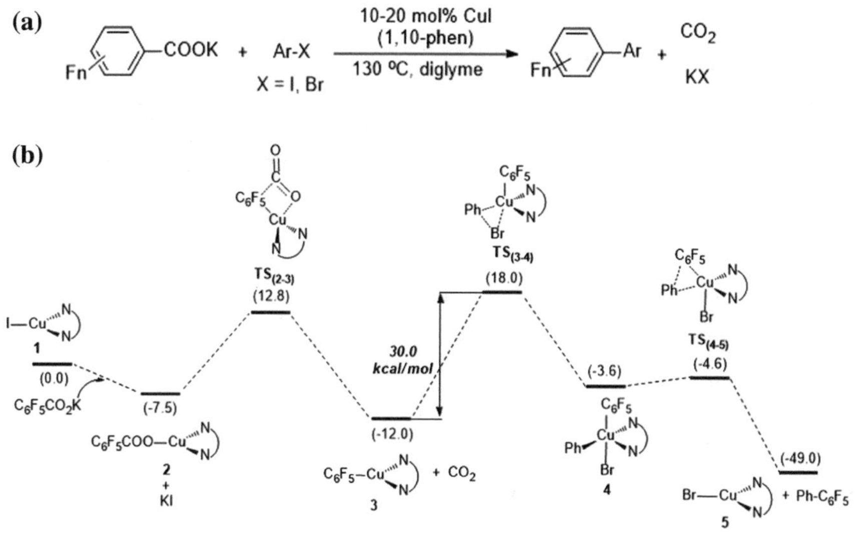

(a)

(b)

Scheme 1.26 a Cu-catalyzed decarboxylative coupling of potassium polyfluorobenzoates with aryl halides, **b** Mechanism (DFT) study on Cu-catalyzed decarboxylative coupling of potassium polyfluorobenzoates with aryl halides

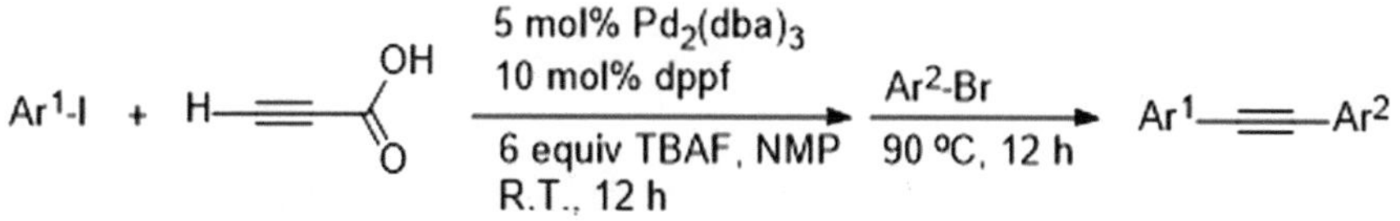

Scheme 1.27 One-pot synthesis of diarylalkynes using Pd-catalyzed Sonogashira reaction and decarboxylative coupling

1.3.1.2 Aryl Alkyne Formation

Lee et al. described that aryl propiolic acid can be formed though propiolic acid with aryl iodides and bromides via Sonogashira coupling reaction, and then, the Pd-catalyzed decarboxylative cross-coupling of aryl propiolic acid with electrophilic reagent occurs to produce nonsymmetric alkynes (Scheme 1.27) [34].

By optimizing the ligand, Li et al. extended the Pd-catalyzed decarboxylative cross-coupling reaction of aryl iodides and bromides to aryl chlorides and benzyl halides (Scheme 1.28) [35].

In 2010 You et al. also reported that the decarboxylative cross-coupling of alkynyl carboxylic acids with aryl bromides and iodides can be accomplished using the Cu(I) catalyst and 10% 1,10-Phen to form nonsymmetric alkynes (Scheme 1.29) [36].

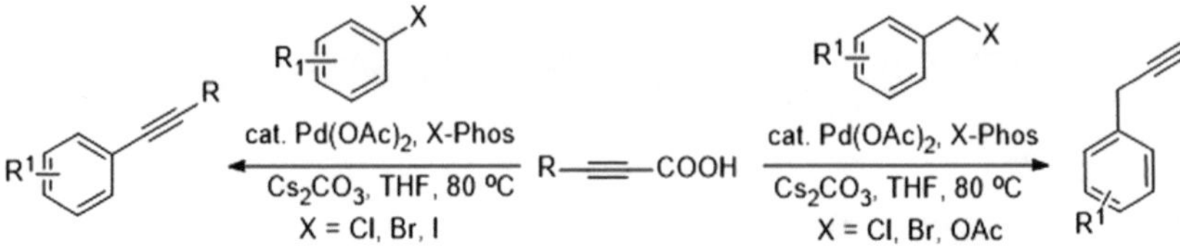

Scheme 1.28 Decarboxylative coupling of alkynyl carboxylic acids with aryl or benzyl halids

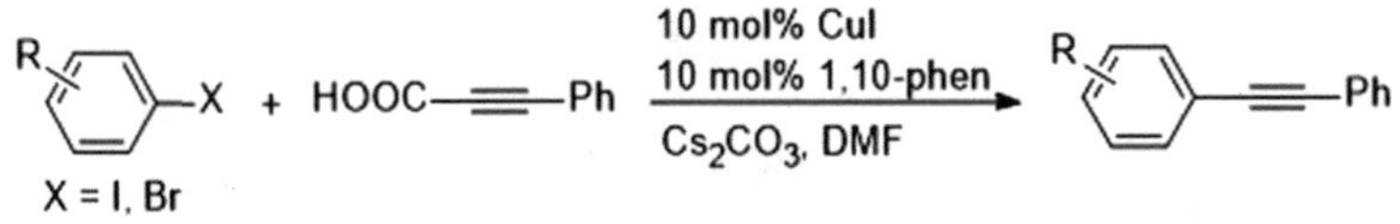

Scheme 1.29 Cu-catalyzed decarboxylative coupling of alkynyl carboxylic acids with aryl halids

1.3.1.3 Aryl Ketone Formation

Goossen et al. extended the concept of decarboxylative cross-coupling to 2-ketone carboxylic acids (Scheme 1.30) [37]. An acyl-anion equivalent is generated in the reaction, and the product is an aryl ketone. Note that this reaction involves a Pd/Cu bimetallic catalyst system.

1.3.1.4 Aryl Ester Formation

Liu et al. reported the Pd-catalyzed decarboxylative cross-coupling of aryl iodides, bromides, and chlorides with potassium oxalate monoesters [38]. This method only needs 1% Pd catalyst and 1.5% dppp (with NMP at 150 °C), and it can realize the conversion of the aryl bromides to the aromatic esters with great compatibility of various functional groups. When using DCyPP (an electron-rich ligand with large steric hindrance), aryl electrophilic reagent can be extended to aryl chlorides. Mechanism study pointed out that the aryl halide underwent oxidative addition on the Pd(0) complexes, then aryl-Pd(II) complexes were formed, and then potassium oxalate anions and halide anions would exchange the anions. The acid decarboxylated on the Pd(II) complexes; thus followed by the reductive elimination, the Pd(0) regeneration and the aromatic ester can be obtained as final product. Theoretical studies indicated that potassium oxalate monoesters on dppp aryl palladium complexes need to overcome the energy barrier of 29.5 kcal to decarboxylate, which is corresponding to the reaction temperature. The theoretical calculation results also supported the proposed mechanism (Scheme 1.31).

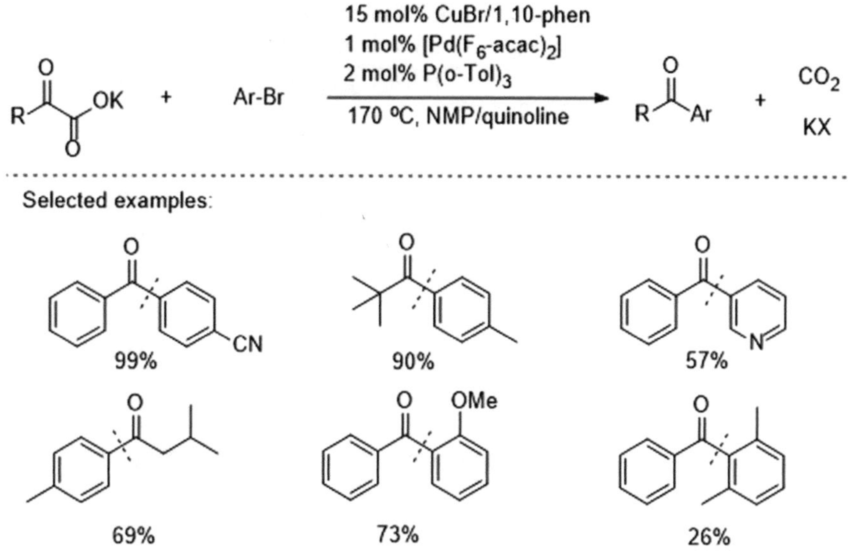

Scheme 1.30 Decarboxylative ketone synthesis reported by Goossen et al.

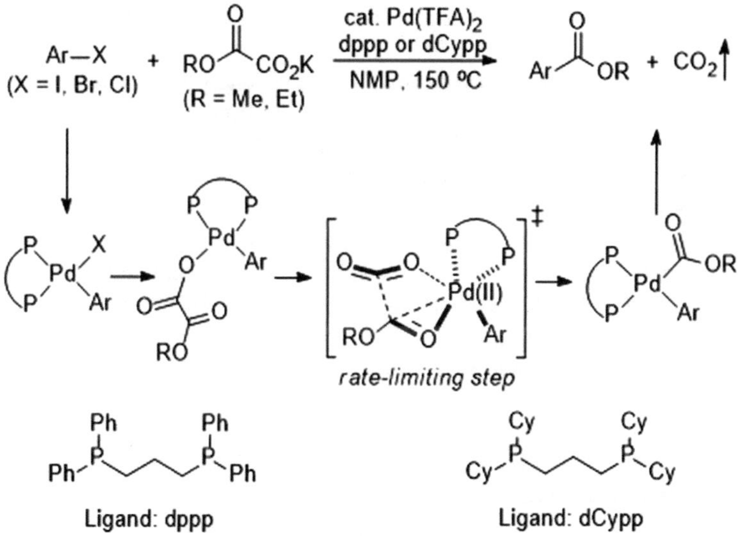

Scheme 1.31 Decarboxylative ester synthesis reported by Liu et al.

1.3.1.5 Decarboxylative Cross-Coupling of C(sp³)–COOH and C–X

Although a lot of studies were related to palladium-catalyzed decarboxylative coupling reaction of aryl halides, most of the reactions were limited in aromatic

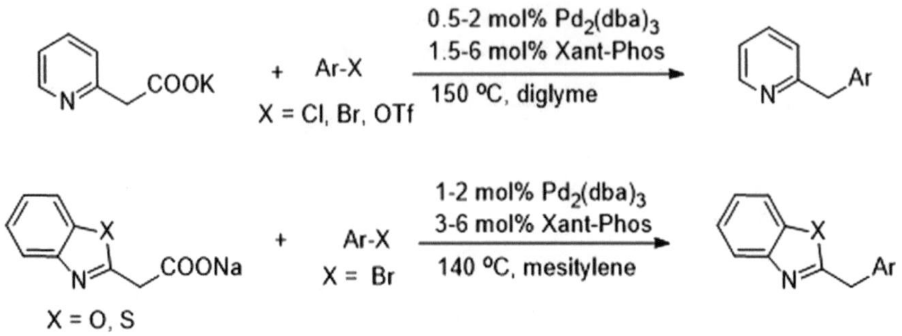

Scheme 1.32 Pd-catalyzed decarboxylative arylation of 2-(2-azaaryl)acetates reported by Liu et al.

carboxylic acids, alkenyl and alkynyl carboxylic acids. Decarboxylative coupling reaction for $C(sp^3)$–COOH carboxylic acid has not been reported. In 2010 Liu et al. reported the first examples of the Pd-catalyzed decarboxylative cross-couplings of 2-(2-azaaryl) acetates with aryl halides and triflates (Scheme 1.32) [39]. This reaction is potentially useful for the synthesis of functionalized pyridines, quinolines, pyrazines, benzoxazoles, and benzothiazoles. This observation is consistent with the theoretical analysis that the nitrogen atom at the 2-position of the heteroaromatics directly coordinates to Pd(II) in the decarboxylation transition state (Scheme 1.33).

Recently, Liu et al. showed that Pd can catalyze the decarboxylative cross-coupling of a cyanoacetate salt and its substituted derivatives with aryl chlorides, bromides, and even triflates (Scheme 1.34) [40]. Note that cyanoacetate salts bearing one or two α-substituents are allowed in the reaction, so that the reaction is potentially useful for the preparation of diverse α-aryl nitriles from readily accessible reactants. The authors also described the synthesis of 2-phenyl acetate using the decarboxylative cross-couplings of a potassium malonic acid monoester. Decarboxylation α-arylation reaction avoids the use of the strong base, which is air sensitive and expensive. The absence of strong base in the reaction system avoids the undesired side reaction and renders better functional group compatibility (Scheme 1.35).

1.3.1.6 Other Examples

Miura et al. reported Pd-catalyzed decarboxylative cross-coupling of alkenyl carboxylic acids with vinyl bromides (Scheme 1.36) [41]. This reaction can be used for the preparation of conjugated dienes.

Pd-catalyzed decarboxylative cross-coupling of aromatic carboxylic acids with allyl halides has also been reported, and this reaction can be used for synthesizing some of the allyl-substituted aromatic hydrocarbon derivatives (Scheme 1.37) [42].

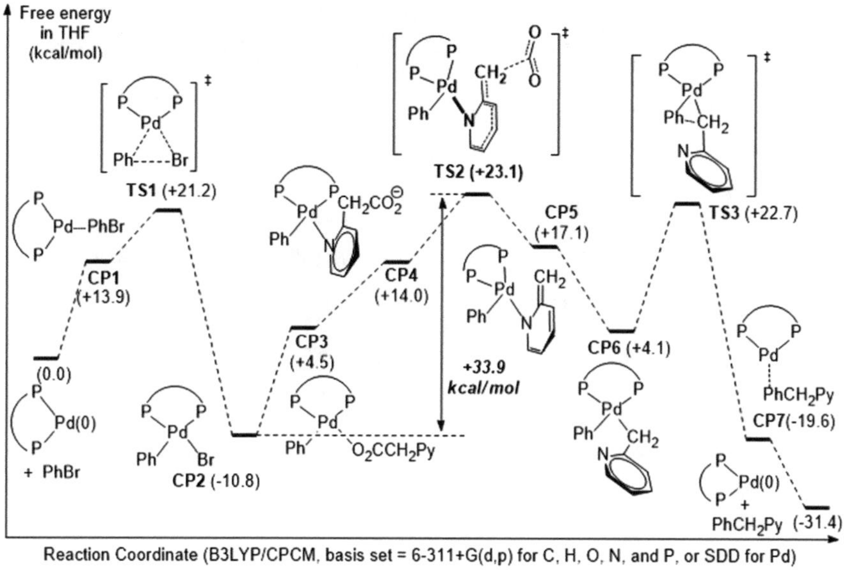

Reaction Coordinate (B3LYP/CPCM, basis set = 6-311+G(d,p) for C, H, O, N, and P, or SDD for Pd)

Scheme 1.33 Mechanistic (DFT) study on Pd-catalyzed decarboxylative coupling of potassium 2-(2-azaaryl)acetates with aryl halides

Liu et al. found a very intriguing example for the decarboxylative cross-coupling of pentafluorobenzoic acid with 1-bromoadamantane (Scheme 1.38) [33].

1.3.2 Decarboxylative Heck Couplings

In 2002 Myers et al. reported decarboxylative Heck reaction of *ortho*-substituted arene carboxylates in a mixed solvent composed of DMF and DMSO (v/v = 20:1) [43]. This reaction is different from the above decarboxylative cross-coupling reactions because it requires three equiv of oxidant such as Ag_2CO_3. Myers et al. had drawn a conclusion that DMSO served as a ligand of Pd, and this effect could make the reaction faster (Scheme 1.39).

Mechanistic studies showed that Pd(II)-mediated decarboxylative palladation constitutes the first step of the catalytic cycle. This step generates an aryl-Pd intermediate, which can react with an alkene through olefin insertion followed by β-hydride elimination to produce the aryl alkene product and a Pd(0) complex. Finally, Pd(0) is oxidized to Pd(II) and reenters the catalytic cycle (Scheme 1.40) [44]. However, a limitation of Myer's decarboxylative Heck coupling is the need to use relatively expensive Ag salts as the oxidant. To solve the problem, Su et al. found that benzoquinone and molecular oxygen can also be used as the oxidant (Scheme 1.41) [45, 46].

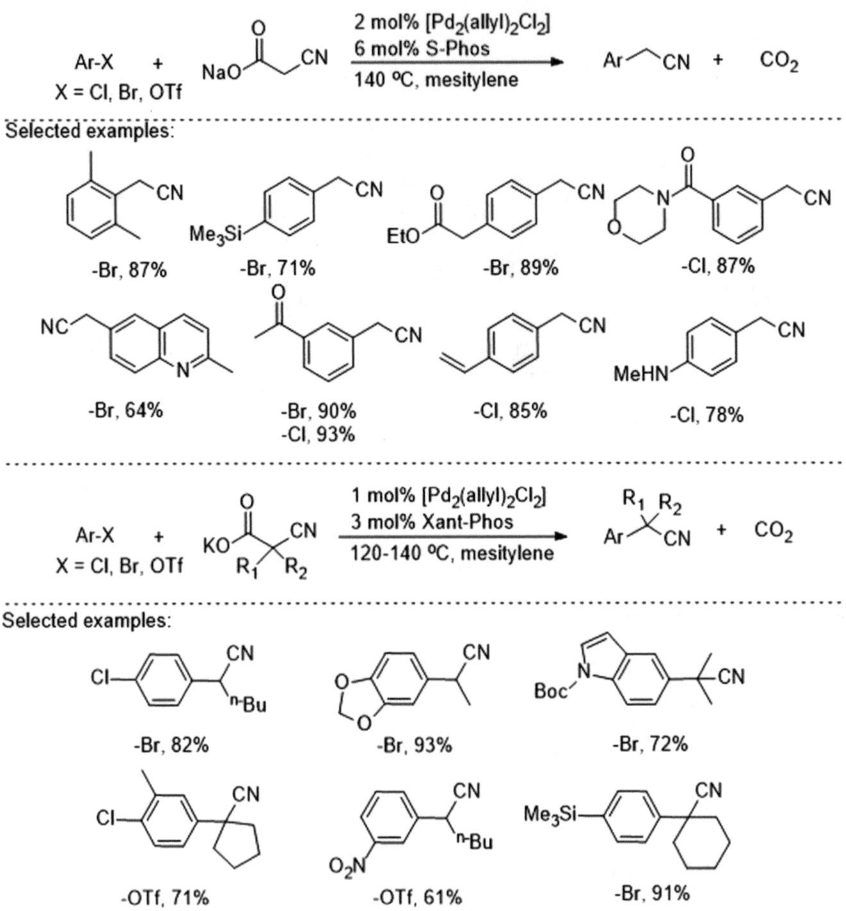

Scheme 1.34 Pd-catalyzed decarboxylative arylation of cyanoacetate salts and derivatives

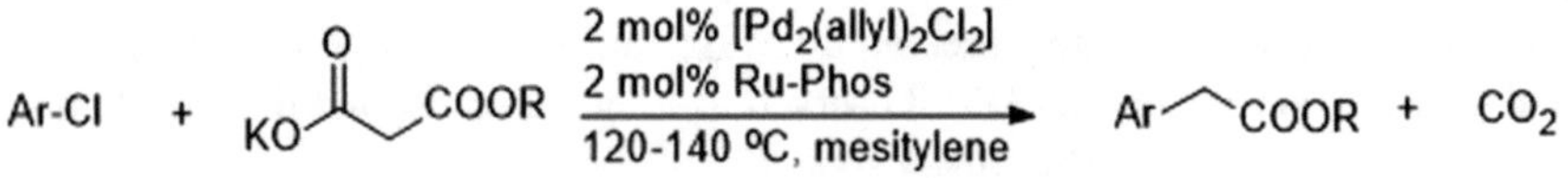

Scheme 1.35 Pd-catalyzed decarboxylative coupling of potassium malonate monoesters with aryl chlorides

Liu et al. researched the reaction mechanism of this decarboxylative Heck reaction with a theoretical chemistry method. They found that the rate-determining step was decarboxylation and formation of the Pd–C bond. Another progress was a DMSO will dissociate from $Pd(DMSO)_2(OTfa)_2$ and the product can get a lower energy (Scheme 1.42) [47].

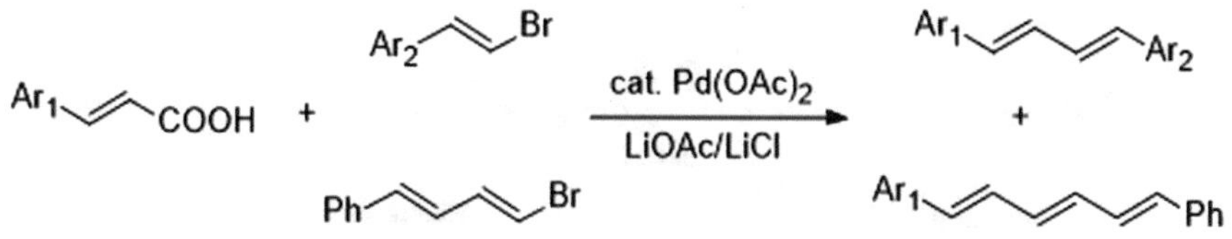

Scheme 1.36 Pd-catalyzed decarboxylative coupling of cinnamic acids with vinyl bromides

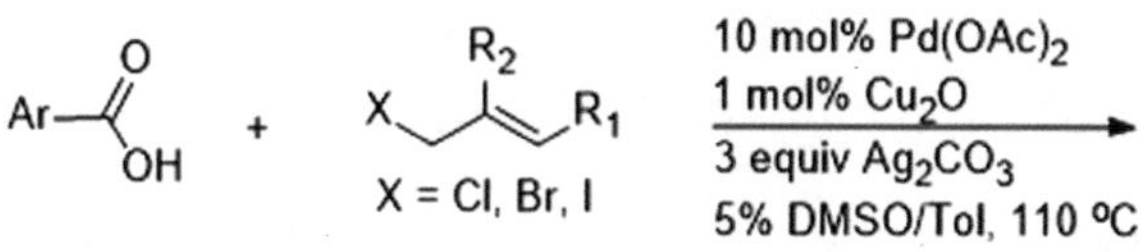

Scheme 1.37 Pd-catalyzed decarboxylative allylation of arenes

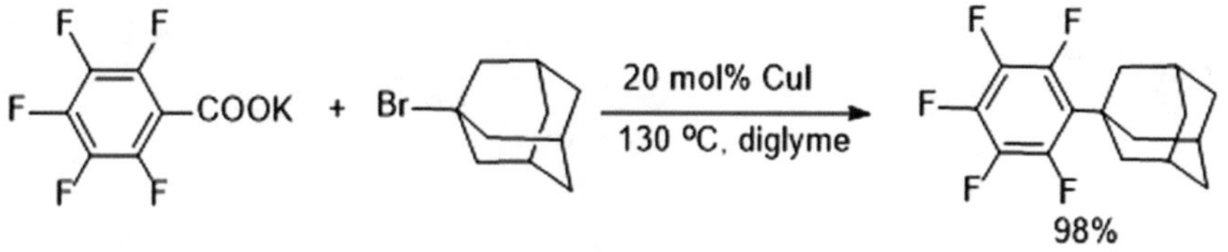

Scheme 1.38 Cu-catalyzed decarboxylative cross-coupling of potassium pentafluorobenzoate with 1-bromoadamantane

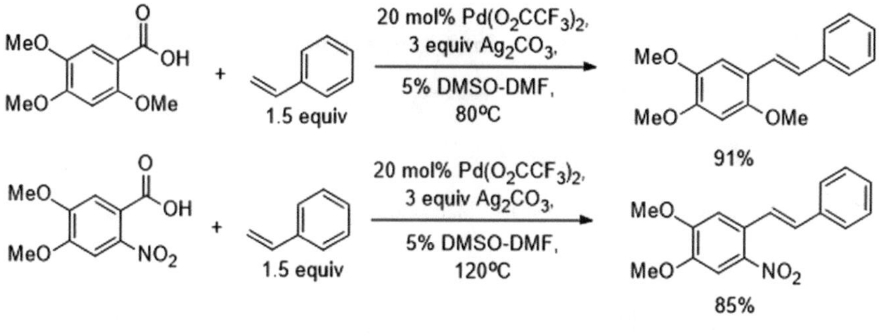

Scheme 1.39 Pd-catalyzed decarboxylative Heck-type olefination of arene carboxylates

Zhao et al. described Rh-catalyzed decarboxylative Heck coupling of aromatic carboxylic acids with acrylate esters [17]. The results showed that only 2,6-difluorinated arene carboxylic acids can afford the desired products in good yields. Furthermore, the ligand plays a key role in determining the chemoselectivity of the reaction. When (R,R)-DIOP is used, the product corresponds to the decarboxylative Heck coupling. When rac-BINAP is used, the product corresponds to the

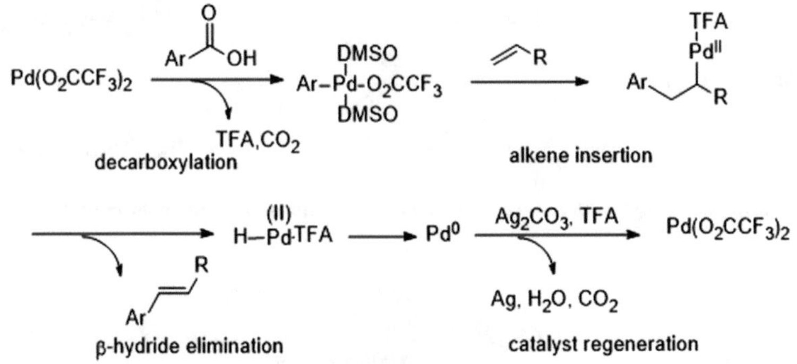

Scheme 1.40 Mechanism of the Pd-catalyzed decarboxylative olefination of arene carboxylic acids

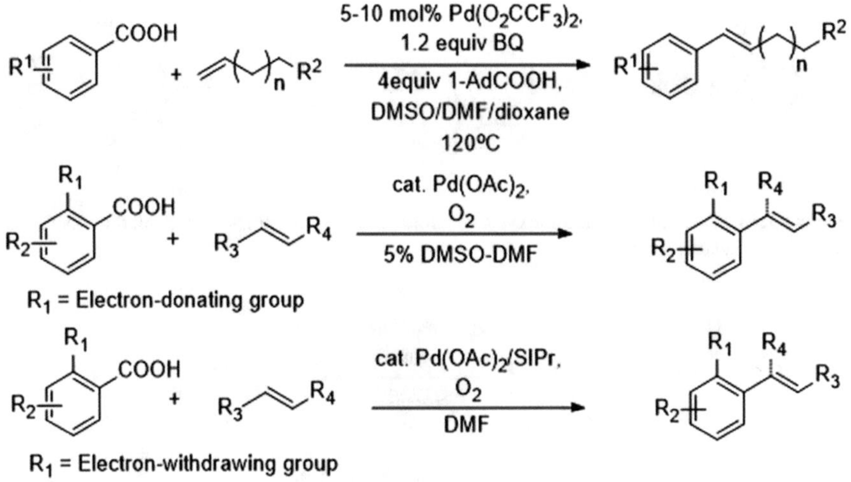

Scheme 1.41 Pd-catalyzed decarboxylative olefination reported by Su and coworkers

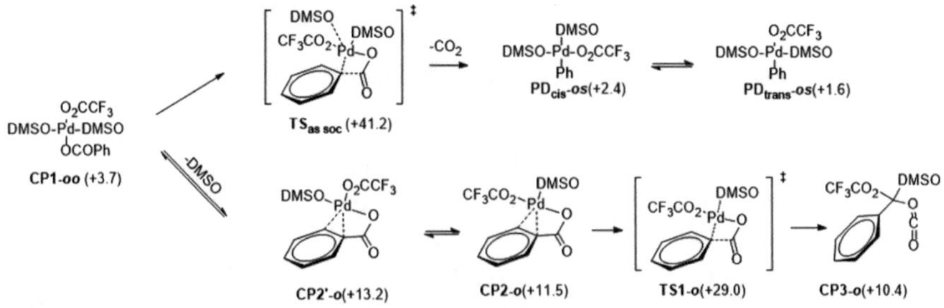

Scheme 1.42 Mechanistic study (DFT) on the decarboxylation step in decarboxylative olefination of arene carboxylic acids

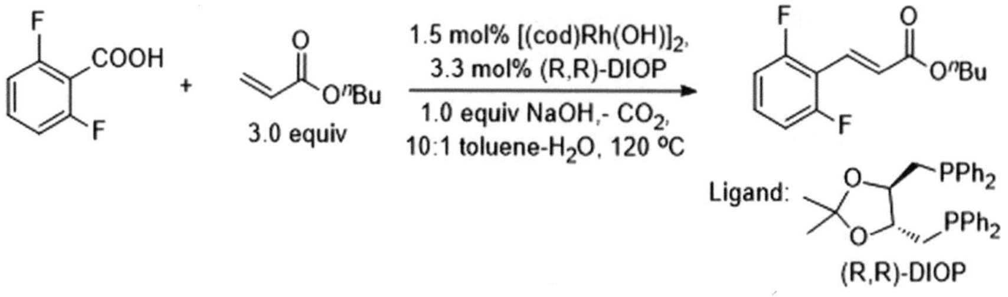

Scheme 1.43 Rh(I)-catalyzed decarboxylative Heck–Mizoroki-type coupling of arenecarboxylic acids

1,4-addition reaction. Note that three equiv of the acrylate ester is used in the reaction, because this substrate also serves as the oxidant (Scheme 1.43).

1.3.3   Decarboxylative Cross-Coupling of Activated C–H/C–COOH

Since the C–H bond is ubiquitous in organic molecules, activation of C–H bond for direct cross-coupling is a simple and efficient manner for efficient bond formations

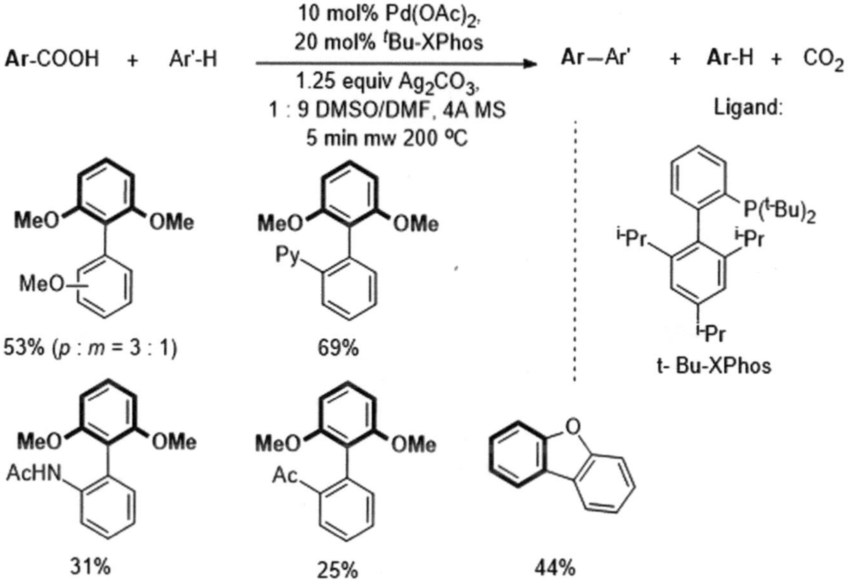

Scheme 1.44 Pd-catalyzed decarboxylative/C–H activate coupling reported by Crabtree and coworkers

that meets the requirement of green chemistry. There has been increasing interest in developing methods for transition metal-catalyzed C–C coupling based on C–H activation. In this context, we discuss recently reported decarboxylative dehydrogenative cross-coupling reactions.

1.3.3.1 Pd-Catalyzed Versions

In 2009 Crabtree et al. reported the first examples of the Pd(II)-catalyzed decarboxylative dehydrogenative cross-coupling of aromatic acids with unactivated arenes using the Buchwald-type ligand [48]. This reaction requires fairly harsh conditions (e.g., microwave heating to 200 °C), but the yield is only modest (36–72%). However, this work points out that decarboxylation reaction can cascade with palladium-catalyzed C–H bond activation, and achieve many useful chemical conversions (Scheme 1.44).

Later Glorius et al. developed an intramolecular decarboxylative dehydrogenative cross-coupling reaction of 2-phenoxybenzoic acids [49]. Pd(TFA)$_2$ was used as the catalyst and the reaction afforded dibenzofurans as the product (Scheme 1.45). The authors proposed that decarboxylative metalation occurs first to generate an aryl-Pd intermediate. Through directed, intramolecular C–H activation, the aryl-Pd intermediate is converted to a palladacycle. Reductive elimination then takes place to provide the desired product and Pd(0). Pd(0) is oxidized to Pd(II) by the Ag salts and reenters the catalytic cycle (Scheme 1.46).

Yu et al. developed an N-oriented C–H bond arylation reaction with benzoyl peroxide as the aryl source. In this reaction, pyridyl, oxazolyl, and methoxy-oxime can be used as the N-orienting group. And this C–H activated reaction possesses a good tolerance of various functional groups. The mechanism which the authors proposed is benzoyl peroxide decompose and decarboxylate first, and the product (aryl radical) will react with palladacycle and go through the reductive elimination to get the final product. Benzoyl peroxide also serves as the oxidant (Scheme 1.47) [50].

As to intermolecular decarboxylative dehydrogenative cross-coupling, Su et al. developed reactions occurring at the 2- or 3-C–H bonds of the indole derivatives [51]. Interestingly, for aromatic carboxylic acids carrying electron-donating groups, the cross-coupling takes place at the 2-position of the indole ring. On the other

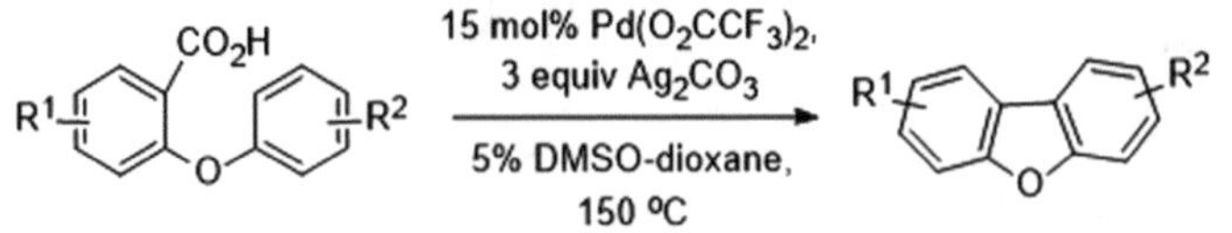

Scheme 1.45 Pd-catalyzed synthesis of dibenzofuran derivatives by tandem decarboxylation/C–H activation

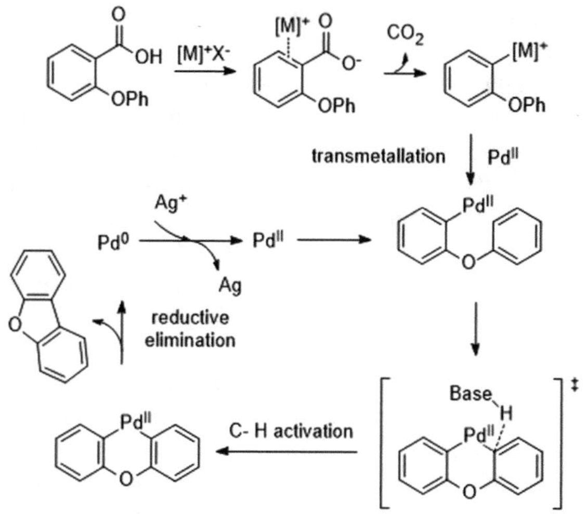

Scheme 1.46 Mechanism proposed by Glorius and coworkers

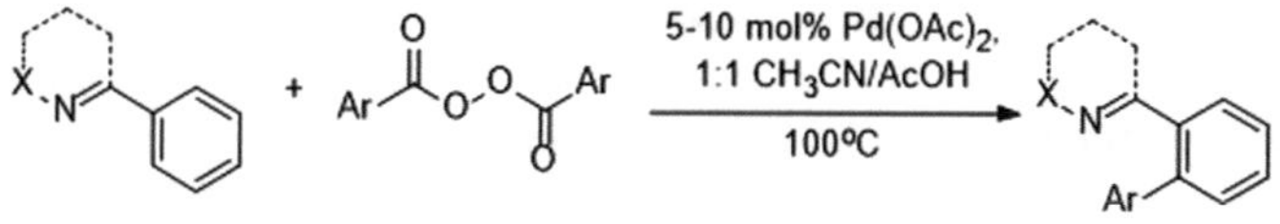

Scheme 1.47 Decarboxylative arylation of aromatic C–H bond using aryl acylperoxides reported by Yu and coworkers

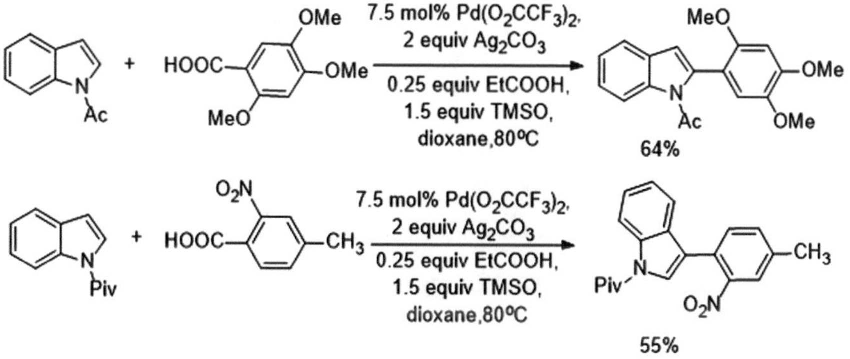

Scheme 1.48 Pd-catalyzed C2/C3 selective arylation of indoles using benzoic acids as arylating reagents

hand, for aromatic carboxylic acids carrying electron-withdrawing groups the major products are 3-arylated indoles (Scheme 1.48).

The author thought that the regioselectivities are determined by different decarboxylation mechanisms. For electron-rich aromatic carboxylic acids Pd catalyzes the decarboxylation (path A), whereas for electron-poor aromatic carboxylic acids Ag promotes the decarboxylation (path B). The transmetalation from aryl-Ag to aryl-Pd may change the regioselectivity (Scheme 1.49).

Larrosa et al. described that intermolecular decarboxylative cross-coupling reaction of electron-poor aromatic carboxylic acids with substituted indoles. Pd $(MeCN)_2Cl_2$ was used as the catalyst and Ag_2CO_3 was used as the oxidant in the reaction (Scheme 1.50) [52].

In 2010, Ge et al. reported the intermolecular decarboxylative dehydrogenative cross-coupling of N-phenylacetamides with 2-ketone acids [53]. This reaction was proposed to proceed through the amide-directed formation of a palladacycle. Surprisingly, the reaction proceeds at room temperature, which contrasts dramatically with most of the decarboxylative cross-coupling reactions. A possible reason is the use of the strong oxidant (i.e., $(NH_4)_2S_2O_8$) in the reaction, which may cause the formation of higher oxidation states of Pd promoting a different type of decarboxylation process (Scheme 1.51).

Nonetheless, Ge et al. also reported the intermolecular decarboxylative dehydrogenative cross-coupling of 2-phenylpyridines with 2-ketone acids. Although $K_2S_2O_8$ was also used in this directed C–H activation reaction, it is surprising that the reaction temperature became 120 °C (Scheme 1.52) [54].

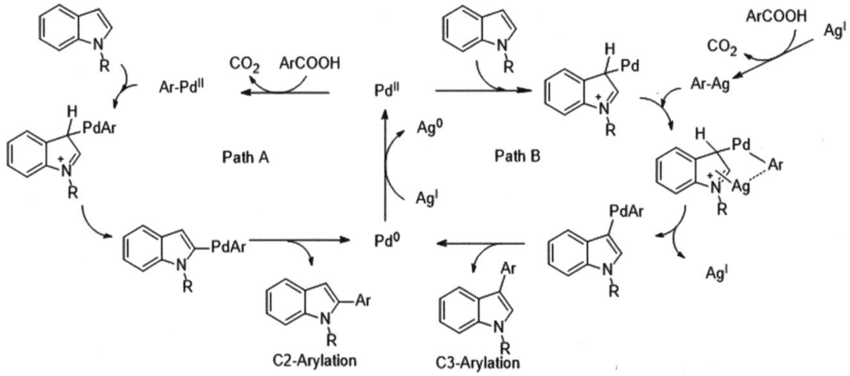

Scheme 1.49 Mechanism and C2/C3 selectivity explained by Su et al.

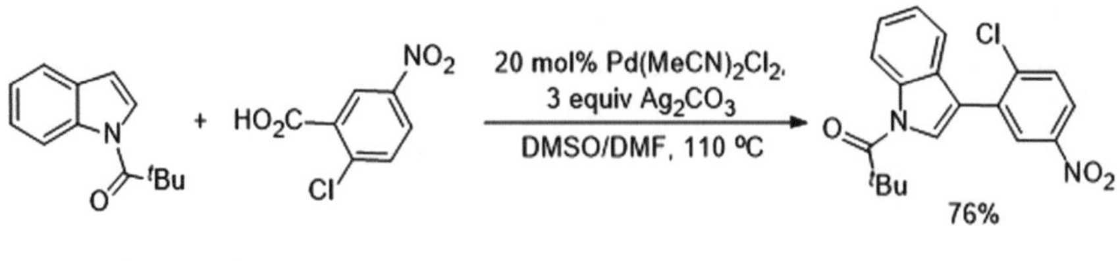

Scheme 1.50 Decarboxylative C3 arylation of indole reported by Larrosa et al.

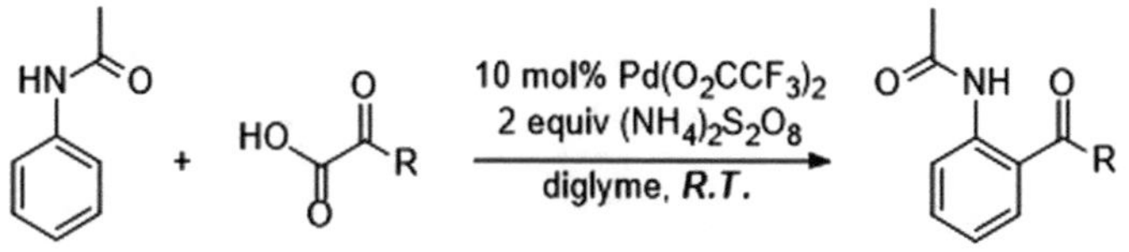

Scheme 1.51 Room temperature Pd-catalyzed decarboxylative ortho-acylation of acetanilides with 2-oxocarboxylic acids

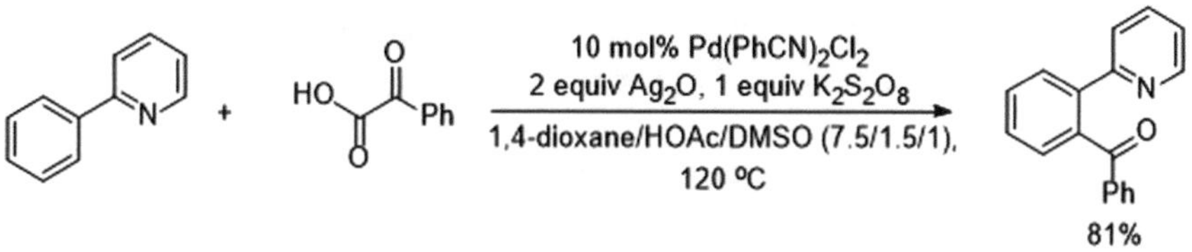

Scheme 1.52 Pd-catalyzed decarboxylative acylation of 2-phenylpyridine with 2-oxocarboxylic acids

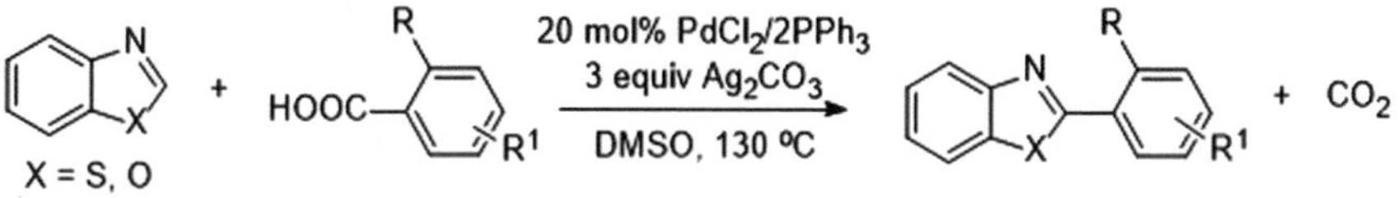

Scheme 1.53 Pd-catalyzed decarboxylative arylation of thiazole and benzoxazole reported by Tan et al.

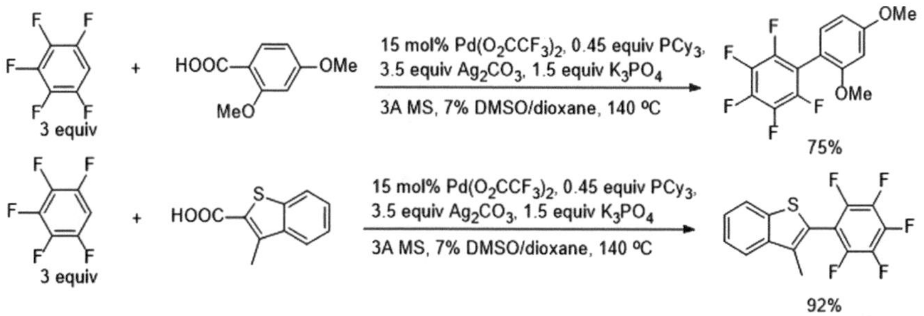

Scheme 1.54 Decarboxylative coupling of arene carboxylic acids with polyfluoroarenes

Tan et al. described the reactions of aromatic carboxylic acids with benzoxazole, benzothiazole, and polyfluorinated benzenes (Scheme 1.53) [55]. Similar reactions were also reported by Su et al. (Scheme 1.54) [56].

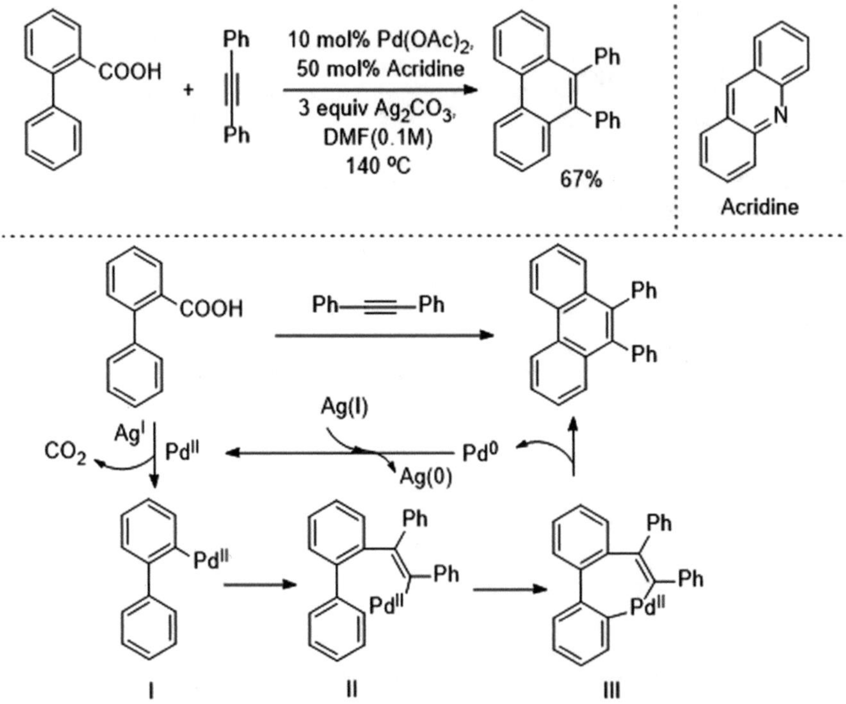

Scheme 1.55 Pd-catalyzed decarboxylative formal [4 + 2] annulation of 2-phenylbenzoic acids with alkynes reported by Glorius et al.

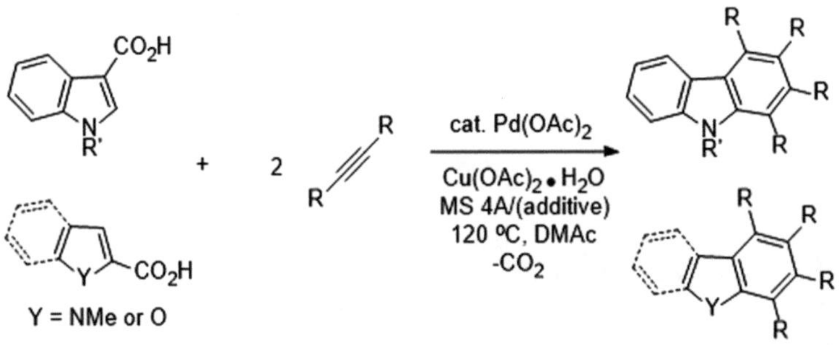

Scheme 1.56 Pd-catalyzed oxidative coupling of heteroarene carboxylic acids with alkynes reported by Miura et al.

To extend the concept of Pd-catalyzed decarboxylative dehydrogenative cross-coupling, Glorius et al. added alkynes into the reaction system. They proposed that 2-phenylbenzoic acid can undergo decarboxylative palladation to generate the aryl-Pd intermediate I, and then the intermediate inserts into the alkyne and is subsequently converted to a palladacycle through intramolecular C–H

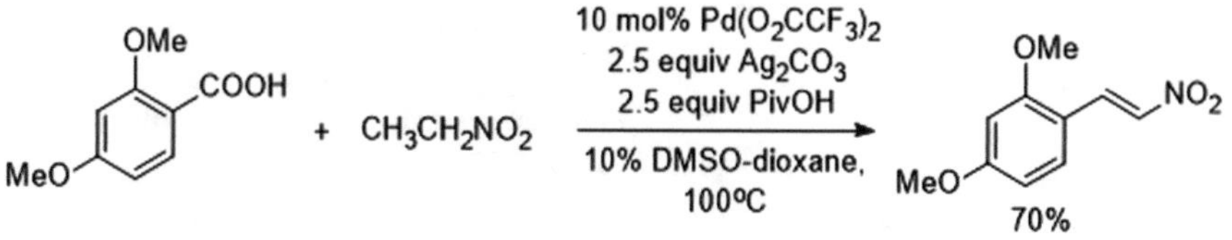

Scheme 1.57 Pd-catalyzed decarboxylative/dehydrogenative couplings of arene carboxylic acids with nitroethane

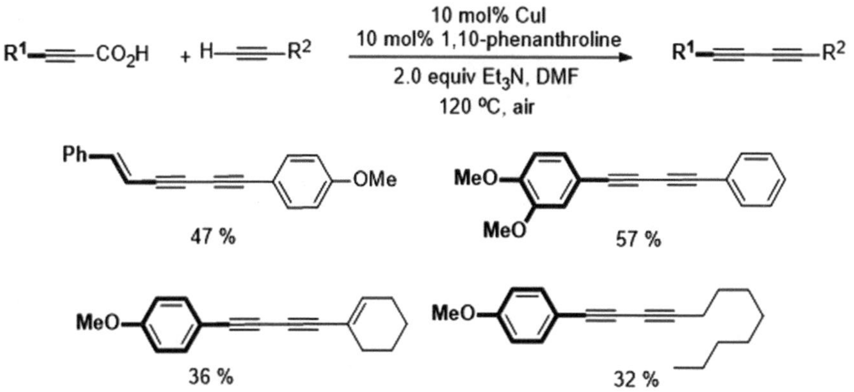

Scheme 1.58 Cu-catalyzed decarboxylative cross-couplings of propiolic acids and terminal alkynes reported by Jiao and coworkers

activation. Finally, reductive elimination takes place to generate phenanthrene as the product (Scheme 1.55) [57].

Miura et al. described decarboxylative dehydrogenative coupling reactions of pyrrole-3-carboxylic acids and furan-3-carboxylic acids with alkynes. Two molecules of alkynes are inserted during the catalytic cycle generating benzopyrroles and benzofurans as the products (Scheme 1.56) [58].

Finally, an interesting reaction was described by Su et al. In the reaction nitroethane reacts with an aromatic carboxylic acid to produce a 2-nitrovinyl arene (Scheme 1.57) [59].

1.3.3.2 Cu-Catalyzed Versions

Much less examples have been reported for Cu-catalyzed decarboxylative dehydrogenative cross-couplings, compared to the Pd-catalyzed versions. For unsaturated carboxylic acids, Jiao et al. once described Cu(I)-catalyzed decarboxylative dehydrogenative cross-coupling of alkynyl carboxylic acids with alkynes. This reaction produces nonsymmetric 1,3-diynes (Scheme 1.58) [60].

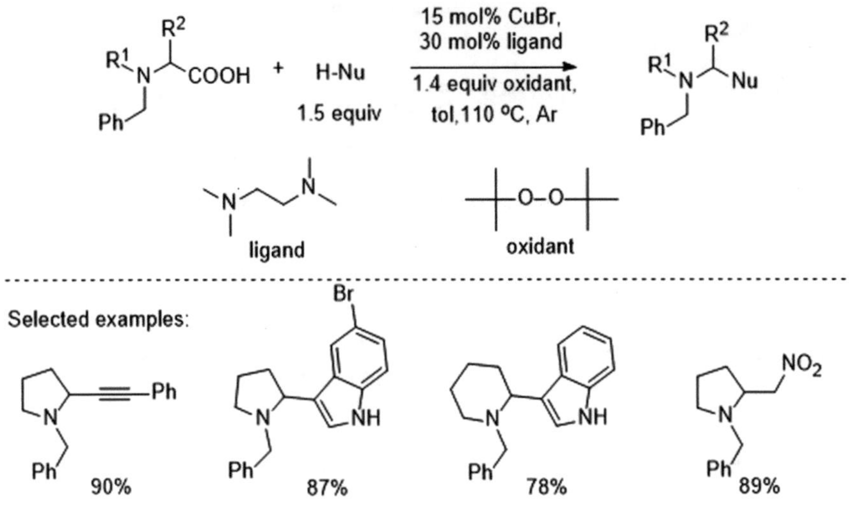

Scheme 1.59 Cu-catalyzed decarboxylative coupling of α-amino acids by Li et al.

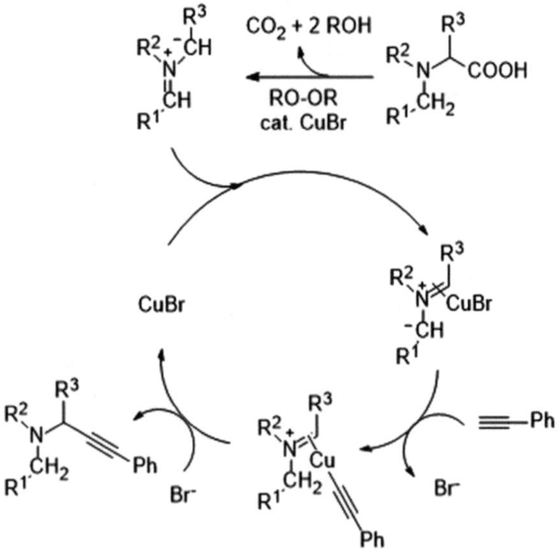

Scheme 1.60 Mechanism proposed by Li et al.

Liang and Li et al. described the Cu-catalyzed cross-coupling of amino acids with alkynes or related nucleophiles. It should be noted that N-alkyl-substituted amino acids is critical to this reaction. The nucleophile reagents may be terminal alkyne, indoles, and nitromethane. The method can be used to synthesize various propargyl amine, indole-3-pyrrolidine, indole-3-piperidine, and β-Nitroamine structure (Scheme 1.59) [61]. This reaction is proposed to involve a

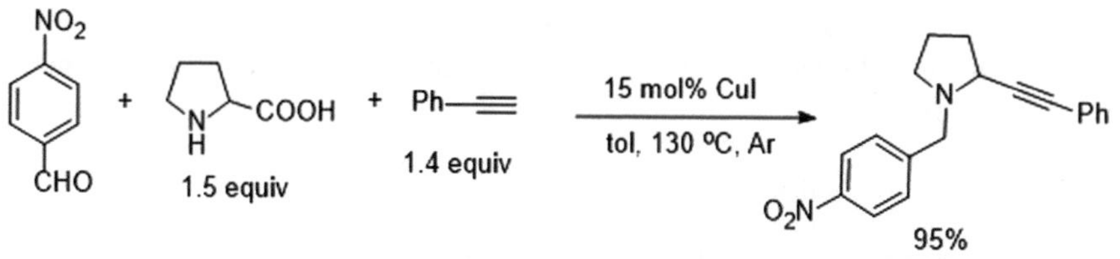

Scheme 1.61 Aldehyde-induced tandem decarboxylation-coupling a-amino acids and alkynes

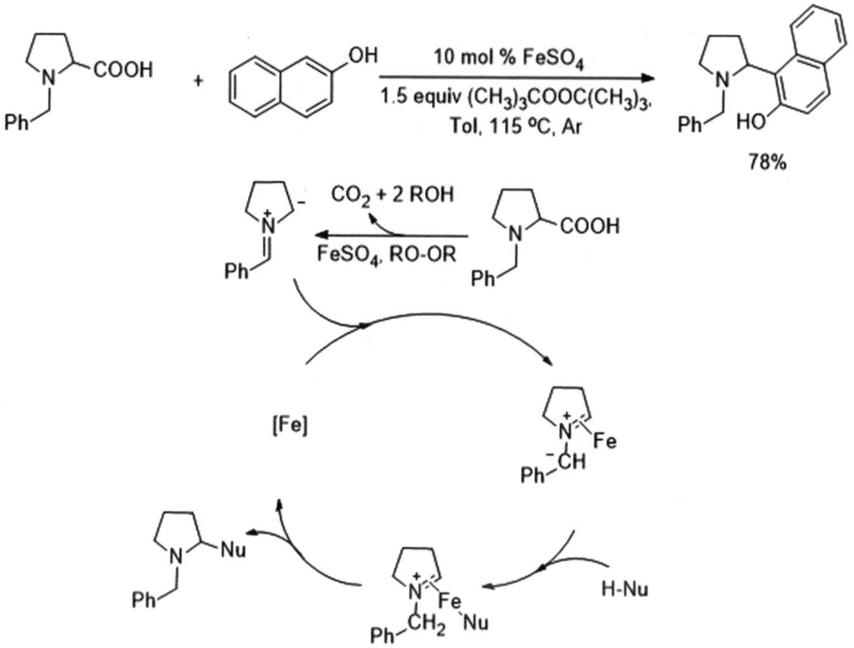

Scheme 1.62 Fe-catalyzed decarboxylative coupling of proline derivatives with naphthol

peroxide-mediated oxidative decarboxylation process. Iminium ions are generated as the intermediates which are then subjected to the attack of nucleophiles to give the target product (Scheme 1.60).

The same authors later showed that the condensations of a carbonyl compound with a proline derivative can also induce the decarboxylation. Subsequent nucleophilic addition again generates the cross-coupling products (Scheme 1.61) [62].

1.3.3.3 Fe-Catalyzed Versions

Liang and Li et al. described that tertiary proline derivatives can be substituted by other nucleophilic reagents (e.g., naphthol) to produce a tertiary amine structure with substituted by naphthol in the presence of Fe(II)-catalyst [63]. The reaction mechanism is similar to the previously reported Cu-catalyzed process (Scheme 1.62).

1.3.4 Cross-Coupling Between C–COOH and C–M Bonds

In addition to the functionalization of C–X and C–H bonds, decarboxylative cross-coupling can also be conducted between a C–COOH moiety and an organometallic reagent.

For example, Liu et al. described decarboxylative cross-coupling of aromatic carboxylic acids with aryl boronic esters. The results showed that the balance between the rate of decarboxylation and transmetalation is critical to the success of the cross-coupling. As a result, different aryl boronic esters must be used in the reactions with acids with different decarboxylation propensities (Scheme 1.63) [64].

Loh et al. described the Pd-catalyzed decarboxylative cross-coupling of alkynyl acids with aryl boronic acids. Ag_2O was used as the oxidant in the reaction and the reaction took place at room temperature (Scheme 1.64) [65].

Recently, Ge et al. reported decarboxylative cross-coupling of trifluoro(organo) borates with 2-ketone acids [66], oxalate monoesters [67], and 2-amino-2-oxoacetates [67] to afford aryl ketones, esters, and amides as the products at room temperature (Scheme 1.65).

1.3.4.1 Cross-Coupling Between C–COOH and C–COOH Bonds

The decarboxylative cross-coupling of two acids is also reported. For example, Larrosa et al. reported the Pd-catalyzed decarboxylative homocoupling between

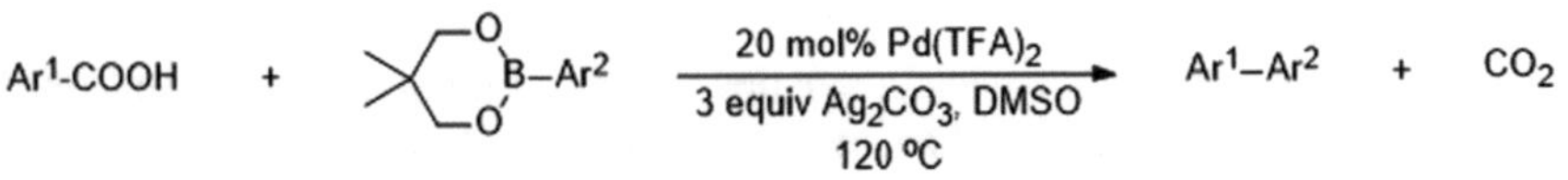

Scheme 1.63 Pd-catalyzed decarboxylative coupling of arene carboxylic acids with aryl boronates

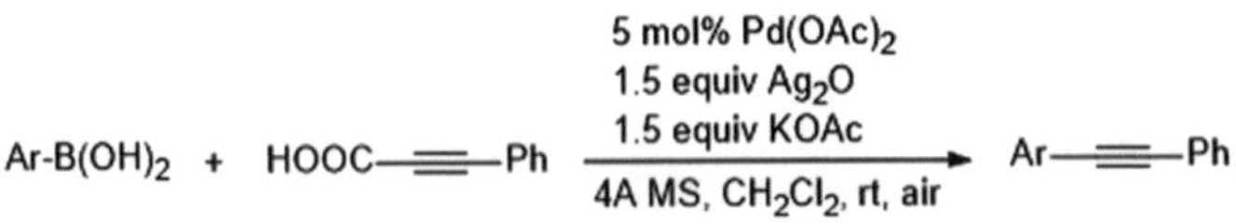

Scheme 1.64 Pd-catalyzed decarboxylative coupling of alkynyl carboxylic acids with aryl boronic acids

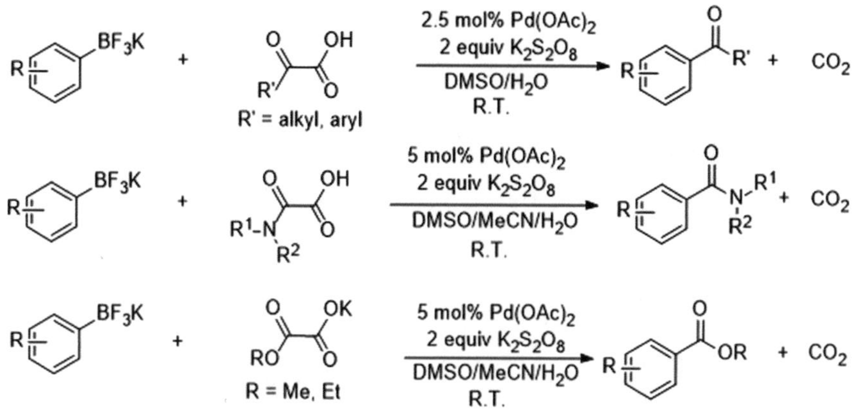

Scheme 1.65 Pd-catalyzed decarboxylative coupling of potassium aryltrifluoroborates with 2-oxocarboxylic acids and oxamic acids reported by Ge et al.

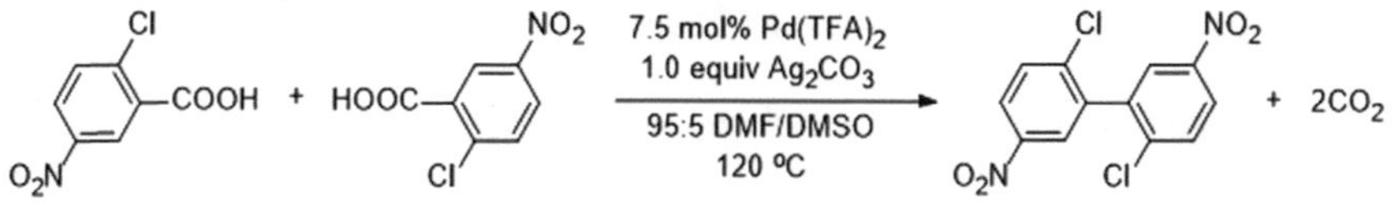

Scheme 1.66 Decarboxylative homocoupling of (hetero) aromatic carboxylic acids reported by Larrosa et al.

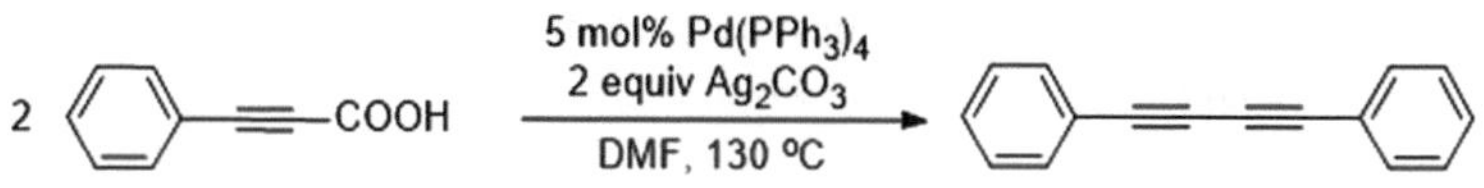

Scheme 1.67 Pd-catalyzed decarboxylative homocouplings of propiolic acids

aromatic carboxylic acids. This reaction is a coupling reaction between two substituent benzoic acid molecules with Pd(TFA)$_2$ and equivalent Ag$_2$CO$_3$. This method can be used in the synthesis of some symmetrical aryl compounds without aryl halides and metal organic reagent (Scheme 1.66) [68].

As we all know, in the Glaser Hay reaction, alkyne acid is easy to decarboxylate. The Pd-catalyzed decarboxylative homocoupling of alkynyl carboxylic acids has also been described (Scheme 1.67) [69].

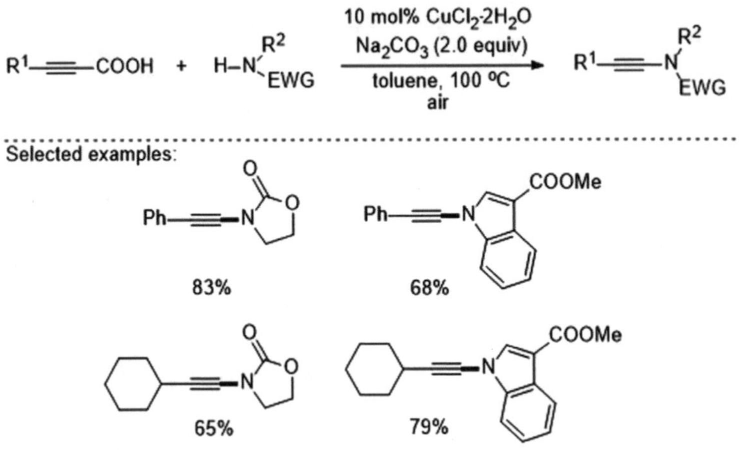

Scheme 1.68 Cu-catalyzed decarboxylative oxidative amidation of propiolic acids

1.4 Decarboxylative Carbon–Heteroatom Cross-Coupling

1.4.1 Decarboxylative C–N Cross-Coupling

Jiao et al. reported the first examples of the Cu-catalyzed decarboxylative C–N cross-coupling of alkynyl carboxylic acids with amides and N-heterocycles (Scheme 1.68) [70]. Air was used as the oxidant in the reaction. This reaction is also a decarboxylative dehydrogenative cross-coupling. Noteworthily, except for this example, no report is available for the decarboxylative C–N cross-coupling that breaks a C_{sp^2}–COOH or C_{sp^3}–COOH bond.

1.4.2 Decarboxylative C–S Cross-Coupling

Liu et al. described an example for catalytic decarboxylative C–S cross-coupling of electron-deficient aromatic acids and heteroaromatic acids with thiol and disulfide in the presence of palladium acetate as catalyst and stoichiometric amount of copper salt as oxidant [71]. The method can be used to synthesize alkyl aryl thioethers and diaryl thioethers (Scheme 1.69).

Liu et al. also reported the decarboxylative C–S cross-coupling of alkynyl carboxylic acids with thiols [72]. It is interesting that the products of the reaction are vinyl thioethers favored with Z-selectivity (Scheme 1.70).

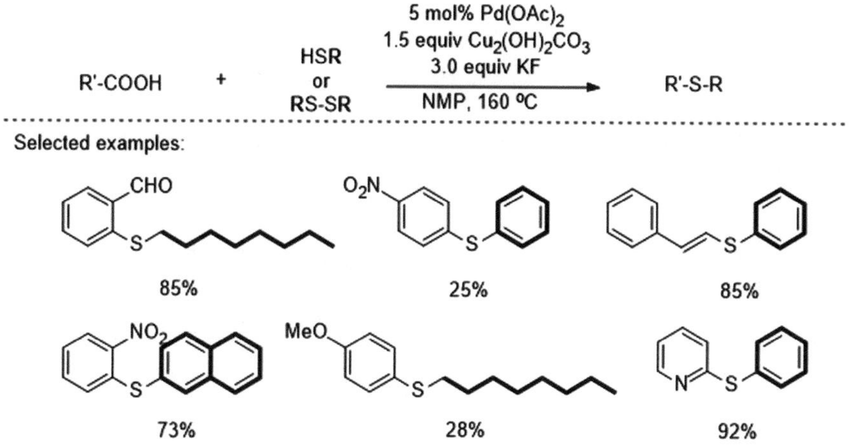

$$R'\text{-COOH} \quad + \quad \begin{array}{c} \text{HSR} \\ \text{or} \\ \text{RS-SR} \end{array} \quad \xrightarrow[\text{NMP, 160 °C}]{\begin{array}{c} \text{5 mol\% Pd(OAc)}_2 \\ \text{1.5 equiv Cu}_2\text{(OH)}_2\text{CO}_3 \\ \text{3.0 equiv KF} \end{array}} \quad R'\text{-S-R}$$

Selected examples:

85%   25%   85%

73%   28%   92%

Scheme 1.69 Synthesis of aryl sulfides via decarboxylative C–S couplings

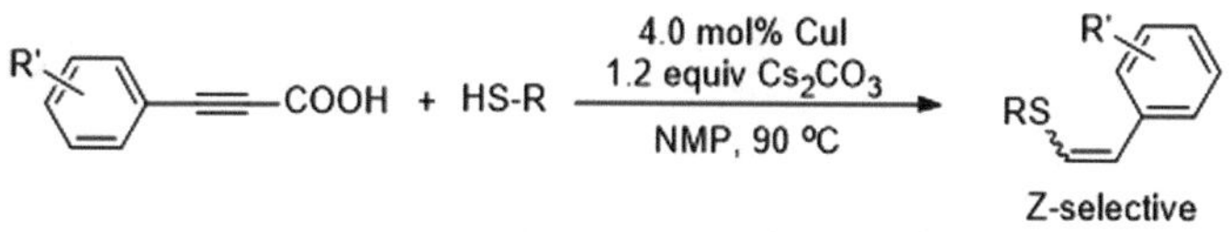

Z-selective

Scheme 1.70 Cu-catalyzed synthesis of vinyl sulfides by decarboxylative coupling of arylpropiolic acids with thiols

1.4.3 Decarboxylative C–P Cross-Coupling

Recently, Yang and Liang et al. reported the first examples of the catalytic decarboxylative C–P cross-coupling of alkynyl and alkenyl carboxylic acids with phosphorous acid derivatives (Scheme 1.71) [73].

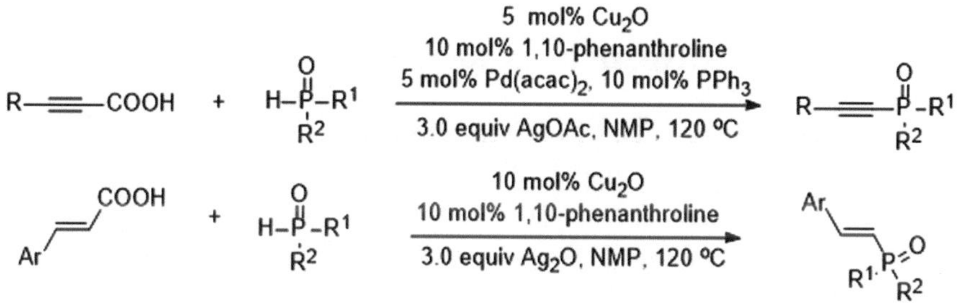

Scheme 1.71 Decarboxylative C_{sp}–P coupling reported by Liang and Yang et al.

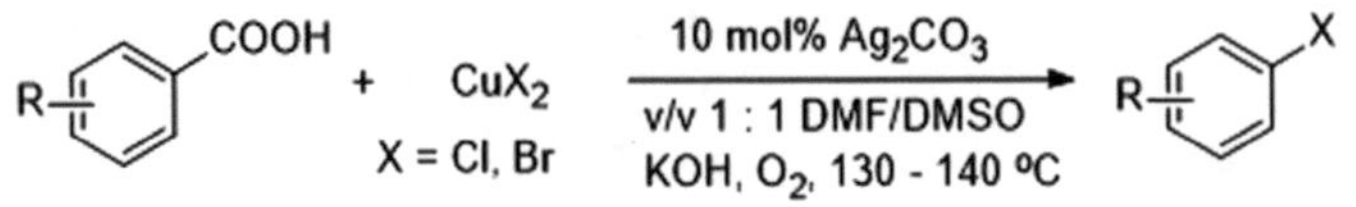

Scheme 1.72 Silver-catalyzed decarboxylative halogenation of carboxylic acids reported by Wu et al.

Scheme 1.73 Halogenation of (hetero)aromatic carboxylic acids via decarboxylative auration

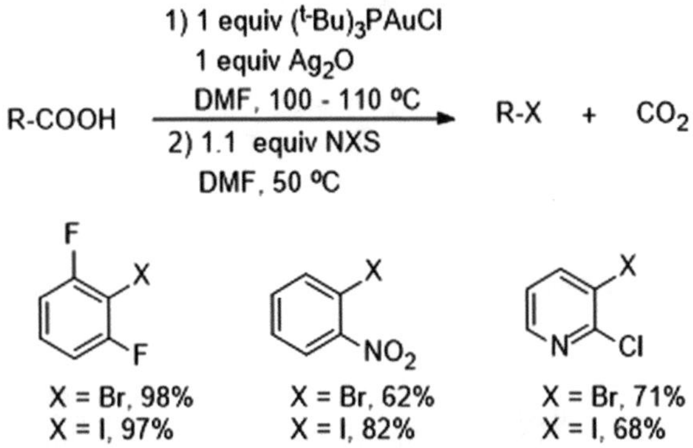

1.4.4 Decarboxylative C-Halide Cross-Coupling

Wu et al. reported a very interesting example for catalytic decarboxylative C–Cl and C–Br cross-couplings of aromatic carboxylic acids [74]. CuCl$_2$ and CuBr$_2$ were used as the reactants, whereas Ag$_2$CO$_3$ was proposed to be the optimal catalyst (Scheme 1.72).

In a related study Larrosa et al. showed that a decarboxylative auration process can be combined with the halogenation reaction using NXS (I, Br). This transformation produces aryl halides from aromatic carboxylic acids. However, a stoichiometric amount of Au and Ag has to be used in the reaction (Scheme 1.73) [75].

1.5 Decarboxylative Addition Reactions

1.5.1 Additions to Aldehydes and Imines

Wu et al. described the first example of Pd-catalyzed decarboxylative 1,2-addition of aromatic carboxylic acids to aldehydes or imines. This reaction was performed under milder conditions, but the substrate scope of the reaction remains limited to 2,6-dimethoxybenzoic acids (Scheme 1.74) [76].

As to Cu catalysis, Shair et al. reported decarboxylative addition of the monothioester of malonic acid to aldehydes [77]. Using chiral ligands, this reaction

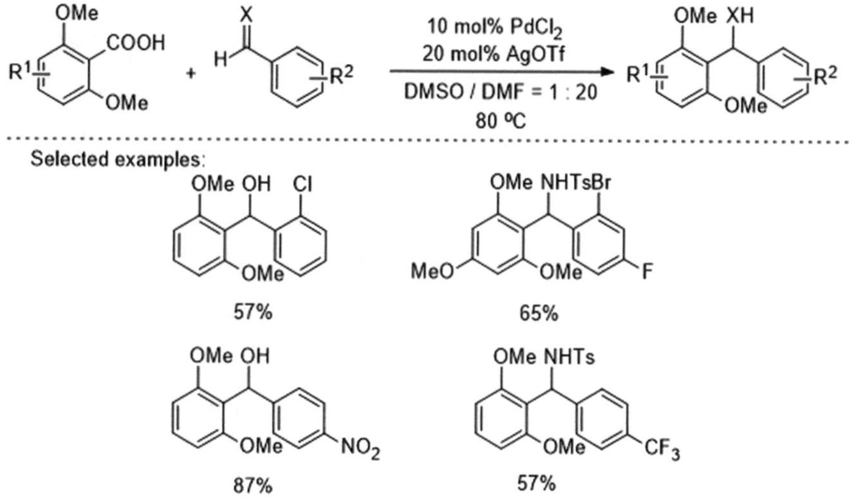

Scheme 1.74 Pd-catalyzed decarboxylative 1,2-addition of carboxylic acids to aldehydes or imines reported by Wu et al.

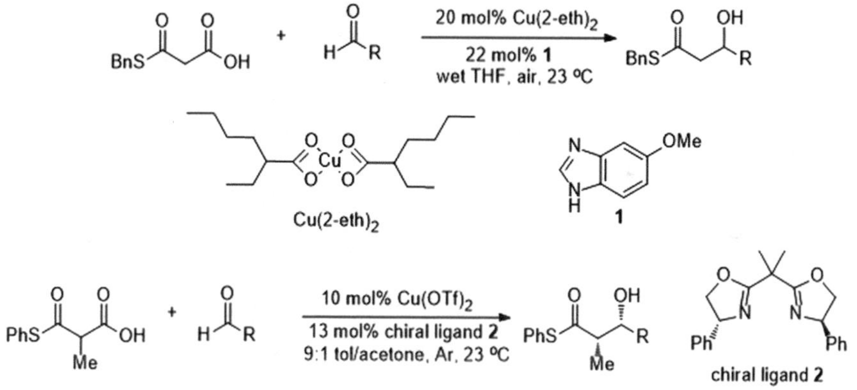

Scheme 1.75 Mild catalytic decarboxylative thioester aldol reaction reported by Shair et al.

can produce β-hydroxy acid derivatives in an enantioselective fashion (Scheme 1.75) [78].

Furthermore, Shibasaki and coworkers reported the Cu-catalyzed enantioselective decarboxylative addition reactions [79]. 2,2-Disubstituted cyanoacetic acids were used as the substrates, which decarboxylated and then were added to imines under the promotion of CuOAc and chiral phosphine ligands. Two chiral centers are generated in the reaction. The reaction shows good diastereoselectivity and enantioselectivity (Scheme 1.76).

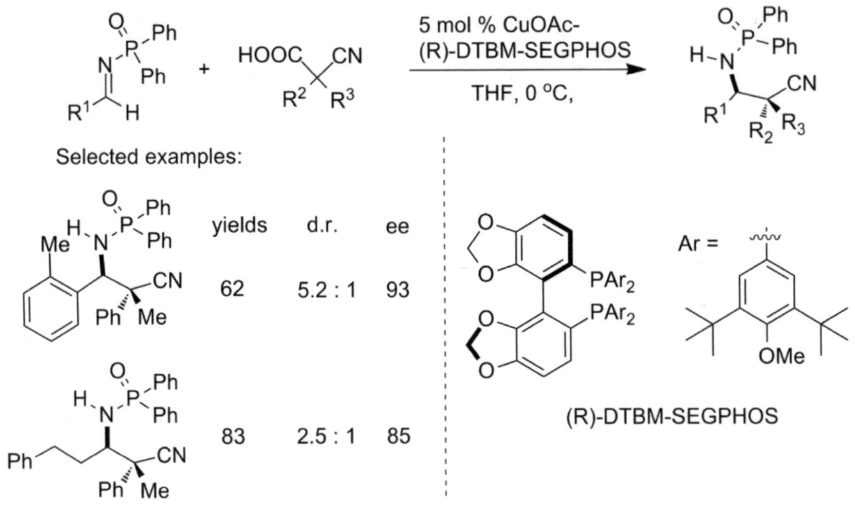

Selected examples:

	yields	d.r.	ee
	62	5.2 : 1	93
	83	2.5 : 1	85

(R)-DTBM-SEGPHOS

Scheme 1.76 Catalytic asymmetric decarboxylative Mannich-type reactions reported by Shibasaki et al.

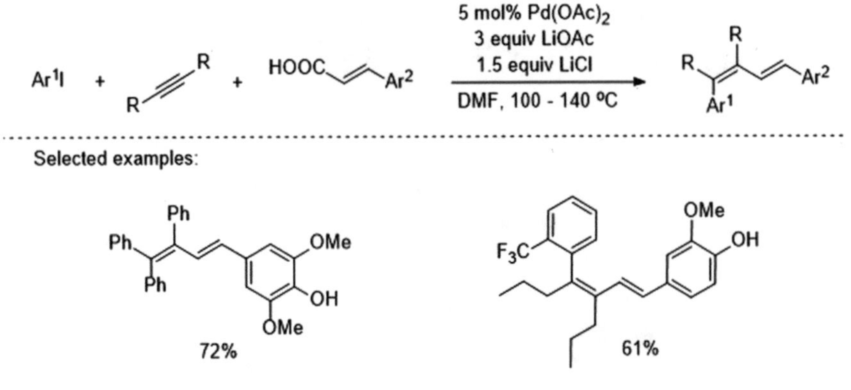

Selected examples:

72%   61%

Scheme 1.77 Synthesis of 1,4-diarylbuta-1,3-dienes through Pd-catalyzed decarboxylative coupling

1.5.2 Additions to Alkynes

Miura et al. reported Pd-catalyzed decarboxylative coupling of unsaturated carboxylic acids with internal alkynes and aryl iodides. Insertion of the decarboxylated intermediate into the alkyne triple bond led to the formation of a 1,4-diarylbuta-1,3-diene as the product (Scheme 1.77) [80].

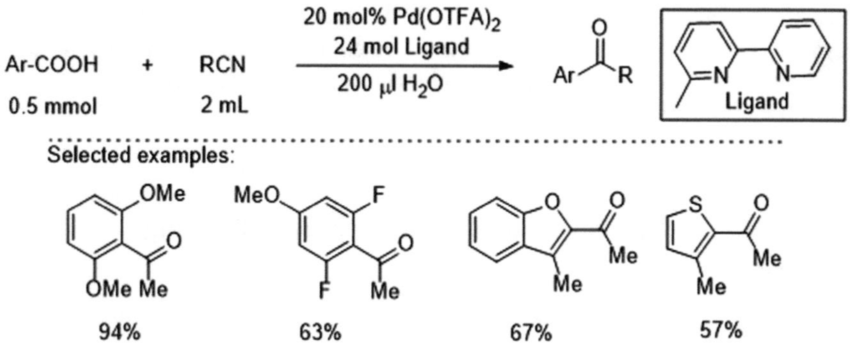

Scheme 1.78 Synthesis of aryl ketones by Pd-catalyzed decarboxylative addition of arene carboxylic acids to nitriles

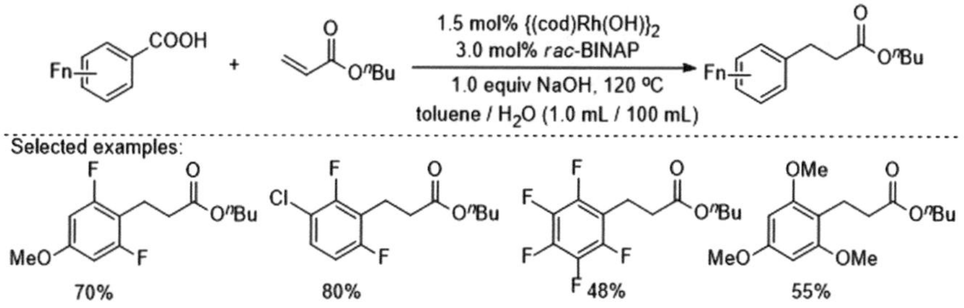

Scheme 1.79 Rh-catalyzed decarboxylative conjugate addition of fluorinated benzoic acids reported by Zhao et al.

1.5.3 Additions to Nitriles

In 2011 Larhed et al. reported Pd-catalyzed decarboxylative addition of aromatic carboxylic acids to nitriles [81]. Imines were proposed as the intermediates, which are hydrolyzed to produce aromatic ketones. The reaction can be applied to 2-methoxybenzoic acids, benzoic acids carrying electron-withdrawing groups, and some heteroaromatic carboxylic acids. A limitation of the reaction is that the nitrile must be used as the solvent (Scheme 1.78).

1.5.4 1,4-Additions

Zhao et al. reported the Rh-catalyzed decarboxylative 1,4-addition of substituted benzoic acids to acrylate esters [82]. The presence of two fluorine atoms at the 2- and 6-positions was found to be necessary for the reaction. The author reported that Rh(I)-catalyzed decarboxylation of a carboxylic acid occurred first to form the aryl-Rh(I) complex, and then, the aryl-Rh(I) complex reacted with acrylic acid derivatives via 1,4-addition. Finally, conjugate addition product was delivered after hydrolysis (Scheme 1.79).

1.6   Other Intramolecular Decarboxylative Coupling Reactions

1.6.1   Decarboxylative Coupling Reaction of Allyl Carboxylates

Some activated allyl carboxylates with electron-withdrawing groups can intramolecularly release carbon dioxide to generate a new C–C bond under Pd catalysis. This transformation was initially reported by Tsuda and Saegusa et al. After that, Tunge, Trost, and Stoltz et al. have greatly extended this approach to different types of substrates [83]. This method can provide access to various sp–sp^3, sp^2–sp^3, sp^3–sp^3 C–C bonds under neutral and mild conditions without using

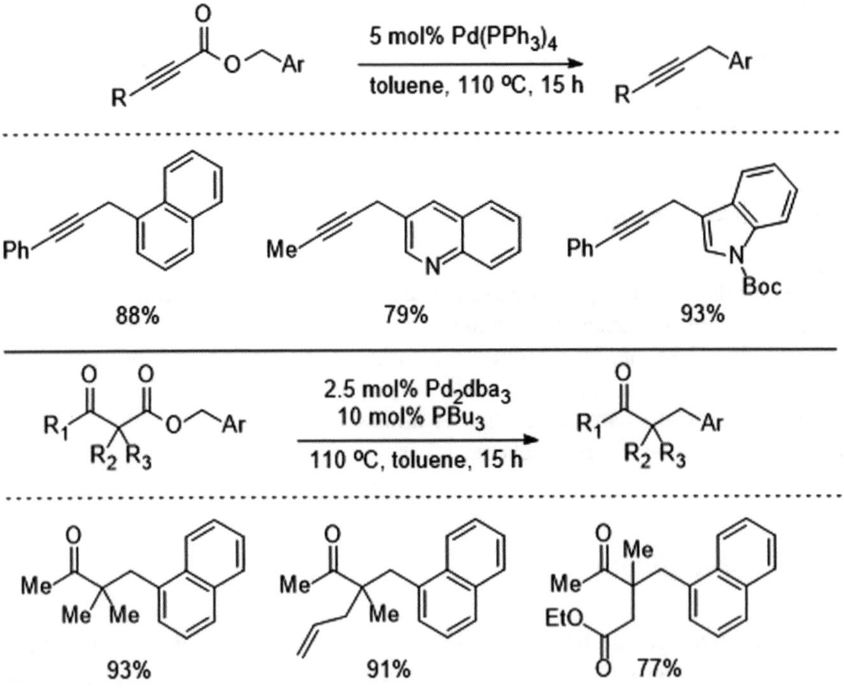

Scheme 1.80 Palladium-catalyzed decarboxylative benzylations of alkynes and ketones

organometallic reagents. Recently, Tunge et al. have reported that various pi-extended benzyl propiolates and benzyl β-ketoesters undergo intramolecular decarboxylative coupling in the presence of a palladium catalyst. The suitable substrates are naphthalen-2-ylmethyl and benzoheterocyclic methyl carboxylate that can achieve a satisfactory yield. Simple benzyl carboxylates are not suitable substrates in this reaction (Scheme 1.80) [84].

1.6.2 Intramolecular Decarboxylative Cyclization

Hayashi et al. described a Pd-catalyzed decarboxylative lactamization of racemic γ-methylidene-δ-valerolactones with isocyanates under mild conditions. Asymmetric lactams can be obtained with high enantioselectivity using this method (Scheme 1.81) [85].

1.7 Application of Decarboxylative Coupling Method in the Synthesis of Important Molecules

Since the decarboxylative coupling reaction methods have many advantages, such as using cheap raw materials that are easy to store and operate, easy to scale up, also environmentally friendly. Several decarboxylarive couplings have been successfully tested in the production of molecules used as pesticides and medicinal compounds.

Goossen et al. developed a new efficient method for synthesis of the angiotensin II inhibitor valsartan through a Cu/Pd bimetallic catalyzed system [86]. In the presence of a catalyst system consisting of copper(II) oxide, 1,10-phenanthroline, and palladium(II) bromide, 2-cyanocarboxylic acid decarboxylatively coupled with

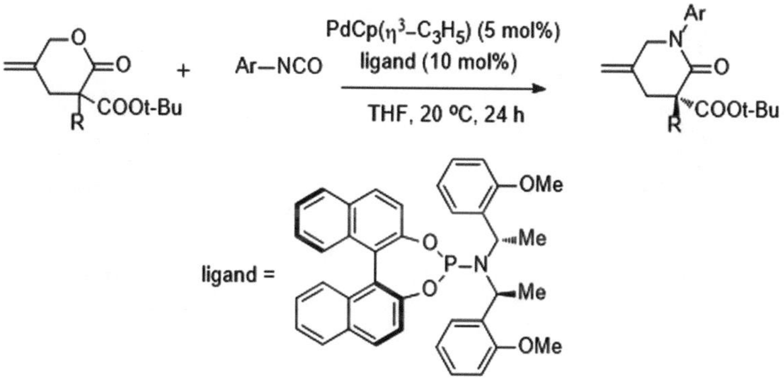

Scheme 1.81 Palladium-catalyzed asymmetric decarboxylative lactamization

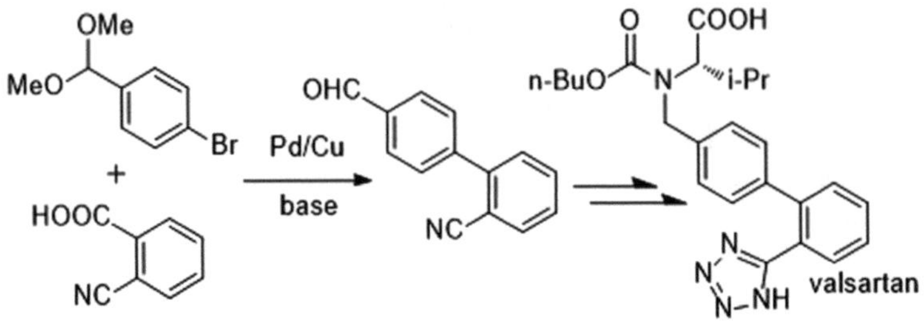

Scheme 1.82 Synthesis of valsartan by Pd/Cu-cocatalyzed decarboxylative coupling reaction

1-bromo (4-dimethoxymethyl)benzene to give 2-cyano-4′-formylbiphenyl. The valsartan synthesis was completed in four steps overall with a total yield of 39%. They pointed out that this method possesses substantial economical and ecological advantages over the traditional process (Scheme 1.82).

Goossen et al. also synthesized angiotensin II receptor antagonist telmisartan using decarboxylative cross-coupling of isopropyl phthalate with 2-(4-chlorophenyl)-1,3-dioxolane as the key step (Scheme 1.83) [87]. Goossen's group also synthesized the key intermediate of Boscalid (a kind of pesticides) using the same method (Scheme 1.84).

Liu et al. applied the decarboxylative coupling method to synthesize the key intermediate of anastrozole, which is a drug used to treat breast cancer, as well as flurbiprofen, which is a nonsteroidal anti-inflammatory drug [88]. As for anastrozole, the decarboxylative coupling of potassium 2-cyano-2-methylpropanoate with 3,5-dibromotoluene affords the key nitrile intermediate in high yield (Scheme 1.85). For flurbiprofen, the decarboxylative coupling of potassium 2-cyanopropanoate with 4-bromo-2-fluorobiphenyl affords the nitrile intermediate, and then hydrolysis of this nitrile intermediate gives the target compound in high yield (Scheme 1.86).

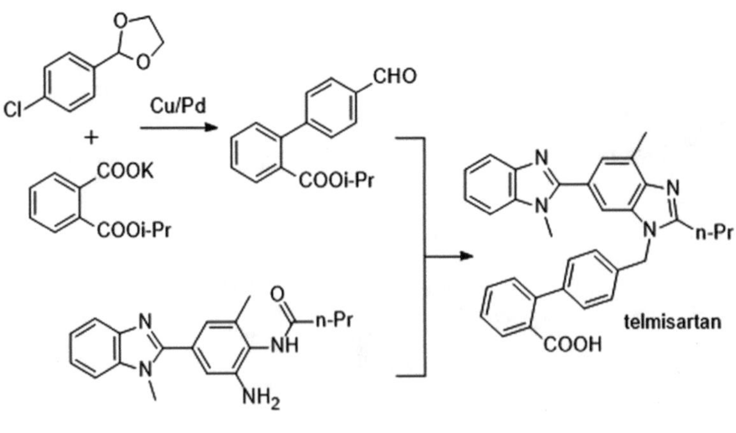

Scheme 1.83 Synthesis of telmisartan by Cu/Pd-cocatalyzed decarboxylative coupling reaction

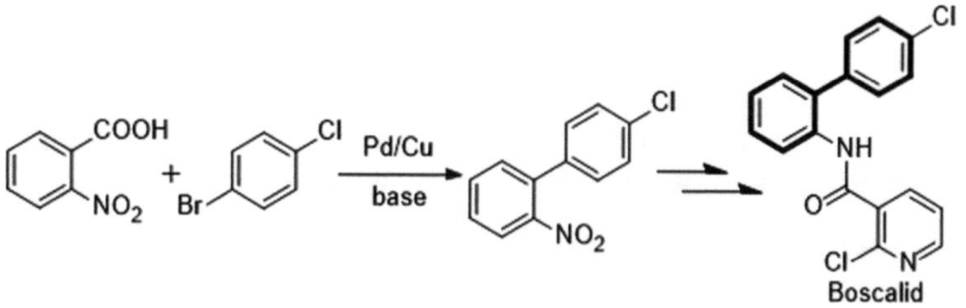

Scheme 1.84 Synthesis of Boscalid by Pd/Cu-cocatalyzed decarboxylative coupling reaction

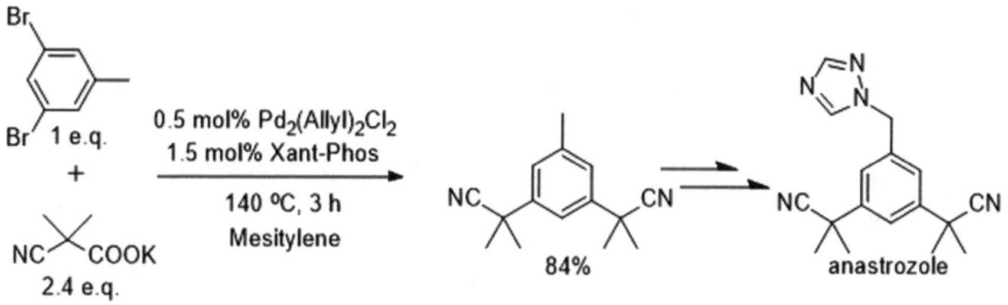

Scheme 1.85 Synthesis of anastrozole by Pd-catalyzed decarboxylative coupling reaction

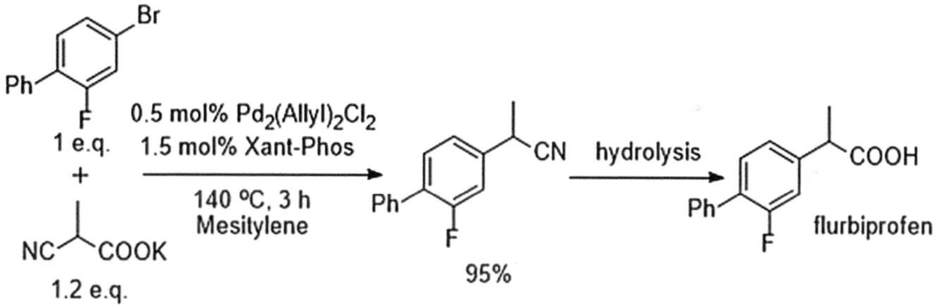

Scheme 1.86 Synthesis of flurbiprofen by Pd-catalyzed decarboxylative coupling reaction

1.8   Conclusions and Perspectives

In summary, we have briefly reviewed transition metal-catalyzed decarboxylative cross-coupling reactions. Transition metal-catalyzed decarboxylative cross-coupling reactions have emerged as a powerful means for the formation of new bonds in organic synthetic chemistry after several decades of development. As shown by the examples in this chapter, catalytic decarboxylative cross-coupling reactions exhibit a number of advantages including the avoidance of using air sensitive or expensive organometallic reagents. They have demonstrated good functional group compatibility, and good chemo- and stereo-selectivities. It is our belief that the transition metal-catalyzed decarboxylative cross-coupling reactions will constitute an important category of organic transformations.

To further expand the scope of transition metal-catalyzed decarboxylative cross-coupling reactions, we need to overcome the following challenges:

1. The substrate scope of the reactions needs to be expanded. More types of aromatic carboxylic acids need to be tested in the reactions. More efforts need to be directed toward the investigation of reactions involving $C_{sp^3}-COOH$.
2. The reaction conditions need to be optimized to meet the demands of various synthetic tasks. For instance, the reaction temperature needs to be lowered and the catalyst loading needs to be reduced.
3. Most of the previous decarboxylative cross-coupling reactions are catalyzed by Pd, Cu, Ag, and Rh. The use of less expensive transition metals like Fe, Ni, and Co in these reactions needs to be studied.
4. New types of decarboxylative cross-coupling reactions need to be developed. The use of these new reactions in the synthesis of natural products or pharmaceutical intermediates needs to be examined.
5. It is important to figure out the detailed mechanism of these reactions, which can guide us to design the new catalysts and reactions.

The above issues are still the remaining challenges and opportunities in organic chemistry research, and this will be an important direction for future research on decarboxylative cross-coupling reactions.

References

1. De Meijere, A., Diederich, F. (2nd Ed.). (2004). Metal-catalyzed Cross-Coupling Reactions: Wiley- VCH, Weinheim.
2. Myers, A. G., Tanaka D., & Mannion, M. R. (2002). *Journal of the American Chemical Society, 124*, 11250–11251.
3. 1) Gooßen, L. J., Deng, G. J., & Levy, L. M. (2006). *Science, 313*, 662–664. 2) Baudoin O. (2007). *Angewandte Chemie International Edition, 46*, 1373–1375.
4. Nilsson, M. (1966). *Acta Chemica Scandinavica, 20*, 423–426.
5. Cohen, T., & Schambach, R. A. (1970). *Journal of the American Chemical Society, 92*, 3189–3190.
6. Cairncross, A., Roland, J. R., Henderson, R. M., & Sheppard, W. A. (1970). *Journal of the American Chemical Society, 92*, 3187–3189.
7. Gooßen, L. J., Thiel, W. R., Rodríguez, N., Linder, C., & Melzer, B. (2007). *Advanced Synthesis & Catalysis, 349*, 2241–2246.
8. Goossen, L. J., Manjolinho, F., Khan, B-A., & Rodríguez, N. (2009). *The Journal of Organic Chemistry, 74*, 2620–2623.
9. Kolarovič, A., & Fáberová, Z. (2009). *The Journal of Organic Chemistry, 74*, 7199–7202.
10. Gooßen, L. J., Linder, C., Rodríguez, N., Lange, P. P., & Fromm, A. (2009). *Chemical Communications*, 7173–7175.

11. Lu, P., Sanchez, C., Cornella, J., & Larrosa, I. (2009). *Organic Letters, 11*, 5710–5713. 2) Cornella, J., Sanchez, C., Banawa, D., & Larrosa, I. (2009). *Chemical Communication*, 7176–7178.
12. Gooßen, L. J., Rodríguez, N., Linder, C., Lange, P. P., & Fromm, A. (2010). *ChemCatChem, 2*, 430–442.
13. Dupuy, S., Lazreg, F., Slawin, A. M. Z., Cazin, C. S. J., & Nolan S. P. (2011). *Chemical Communications, 47*, 5455–5457.
14. Cornella, J., Rosillo-Lopez, M., & Larrosa, I. (2011). *Advanced Synthesis & Catalysis, 353*, 1359–1366.
15. Dickstein, J. S., Mulrooney, C. A., O'Brien, E. M., Morgan, B. J., & Kozlowski, M. C. (2007). *Organic Letters, 9*, 2441–2444.
16. Tanaka, D., Romeril, S. P., & Myers, A. G. (2005). *Journal of the American Chemical Society, 127*, 10323–10333.
17. Sun, Z. M., Zhang, J., & Zhao, P. (2010). *Organic Letters, 12*, 992–995.
18. Rodriguez, N. & Goossen L. J., (2011). *Chemical Society Reviews, 40*, 5030–5048.
19. Goossen, L. J., Zimmermann, B., & Knauber, T. (2008). *Angewandte Chemie International Edition, 47*, 7103–7106.
20. Goossen, L. J., Rodríguez, N., & Linder, C. (2008). *Journal of the American Chemical Society, 130*, 15248–15249.
21. Goossen, L. J., Rodríguez, N., Lange, P. P., & Linder, C. (2010). *Angewandte Chemie International Edition, 49*, 1111–1114.
22. Goossen, L. J., Rodríguez, N., Melzer, B., Linder, C., Deng, G. J., & Levy, L. M. (2007). *Journal of the American Chemical Society, 129*, 4824–4833.
23. Becht, J. M., Catala, C., Drian, C. L., & Wagner, A. (2007). *Organic Letters, 9*, 1781–1783.
24. Becht, J. M., & Drian, C. L. (2008). *Organic Letters, 10*, 3161–3164.
25. Wang, Z.-Y., Ding, Q.-P., He, X.-D., & Wu, J. (2009). *Tetrahedron Letters, 65*, 4635–4638.
26. Zhang, F., & Greaney, M. F. (2010). *Organic Letters, 12*, 4745–4747.
27. Goossen, L. J., Lange, P. P., Rodríguez, N., & Linder, C. (2010). *Chemistry - A European Journal, 16*, 3906–3909.
28. Peschko, C., Winklhofer, C., & Steglich, W. (2000). *Chemistry - A European Journal, 6*, 1147–1152.
29. Forgione, P., Brochu, M. C., St-Onge, M., & Thesen, K. H. (2006). *Journal of the American Chemical Society, 128*, 11350–11351.
30. Shang, R., Xu, Q., Jiang, Y.-Y., Wang, Y., & Liu, L. (2010). *Organic Letters, 12*, 1000–1003.
31. Miyasaka, M., Fukushima, A., Satoh, T., Hirano, K., & Miura, M. (2009). *Chemistry - A European Journal, 15*, 3674–3677.
32. Arroyave, F. A., & Reynolds, J. R. (2010). *Organic Letters, 12*, 1328–1331.
33. Shang, R., Fu, Y., Wang, Y., Xu, Q., Yu, H.-Z., & Liu, L. (2009). *Angewandte Chemie International Edition, 48*, 9350–9354.
34. Moon, J., Jeong, M., Nam, H., Ju, J., Moon, J. H., Jung, H. M., & Lee, S. (2008). *Organic Letters, 10*, 945–948.
35. Zhang, W. W., Zhang, X.-G., & Li, J.-H. (2010). *The Journal of Organic Chemistry, 75*, 5259–5264.
36. Zhao, D.-B., Gao, C., Su, X.-Y., He, Y.-Q., You, J.-S., & Xue, Y. (2010). *Chemical Communications, 46*, 9049–9051.
37. Goossen, L. J., Rudolphi, F., Oppel, C., & Rodríguez, N. (2008). *Angewandte Chemie International Edition, 47*, 3043–3045.
38. Shang, R., Fu, Y., Li, J.-B., Zhang, S.-L., Guo, Q.-X., & Liu, L. (2009). *Journal of the American Chemical Society, 131*, 5738–5739.
39. Shang, R., Yang, Z.-W., Wang, Y., Zhang, S.-L., & Liu, L. (2010). *Journal of the American Chemical Society, 132*, 14391–14393.
40. Shang, R., Ji, D.-S., Chu, L., Fu, Y., & Liu, L. (2011). *Angewandte Chemie International Edition, 50*, 4470–4474.
41. Yamashita, M., Hirano, K., Satoh, T., & Miura, M. (2010). *Organic Letters, 12*, 592–595.

42. Wang, J.-T., Cui, Z.-L., Zhang, Y.-X., Li, H.-J., Wu, L.-M., & Liu, Z.-Q. *Organic & Biomolecular Chemistry, 9*, 663–666.
43. Nref 2 and Tanaka, D., & Myers, A. G. (2004). *Organic Letters, 6*, 433–436.
44. Tanaka, D., Romeril, S. P., & Myers, A. G. (2005). *Journal of the American Chemical Society, 127*, 10323–10333.
45. Hu, P., Kan, J., Su, W., & Hong, M. (2009). *Organic Letters, 11*, 2341–2344.
46. Fu, Z., Huang, S., Su, W., & Hong, M. (2010). *Organic Letters, 12*, 4992–4995.
47. Zhang, S.-L., Fu, Y., Shan, R., Guo, Q.-X., & Liu, L. (2010). *Journal of the American Chemical Society, 132*, 638–646.
48. Voutchkova, A., Coplin, A., Leadbeater, N-E., & Crabtree, R-H. (2008). *Chemical Communications*, 6312–6314.
49. Wang, C. Y., Piel, I., & Glorius, F. (2009). *Journal of the American Chemical Society, 131*, 4194–4195.
50. Yu, W. Y., Sit, W. N., Zhou, Z., & Chan, A. S-C. (2009). *Organic Letters, 11*, 3174–3177.
51. Zhou, J., Hu, P., Zhang, M., Huang, S.-J., Wang, M., & Su, W.-P. (2010). *Chemistry - A European Journal, 16*, 5876–5881.
52. Cornella, J., Lu, P., & Larrosa, I. (2009). *Organic Letters, 11*, 5506–5509.
53. Fang, P., Li, M.-Z., & Ge, H.-B. (2010). *Journal of the American Chemical Society, 132*, 11898–11899.
54. Li, M.-Z., & Ge H.-B. (2010). *Organic Letters, 12*, 3464–3467.
55. Xie, K., Yang, Z., Zhou, X., Li, X., Wang, S., Tan, Z., An, X., & Guo, C.-C. (2010). *Organic Letters, 12*, 1564–1567.
56. Zhao, H.-Q., Wei, Y., Xu, J., Kan, J., Su, W.-P., & Hong, M.-C. (2011). *The Journal of Organic Chemistry, 76*, 882–893.
57. Wang, C. Y., Rakshit, S., & Glorius, F. (2010). *Journal of the American Chemical Society, 132*, 14006–14008.
58. Yamashita, M., Hirano, K., Satoh, T., & Miura, M. (2009). *Organic Letters, 11*, 2337–2340.
59. Zhang, M., Zhou, J., Kan, J., Wang, M., Su, W.-P., & Hong, M.-C. (2010). *Chemical Communications, 46*, 5455–5457.
60. Yu, M., Pan, D.-L., Jia, W., Chen, W., & Jiao, N. (2010). *Tetrahedron Letters, 51*, 1287–1290.
61. Bi, H.-P., Zhao, L., Liang, Y.-M., & Li, C.-J. (2009). *Angewandte Chemie International Edition, 48*, 792–795.
62. Bi, H.-P., Teng, Q.-F., Guan, M., Chen, W.-W., Liang, Y.-M., Yao, X.-J., & Li, C.-J. (2010). *The Journal of Organic Chemistry, 75*, 783–788.
63. Bi, H.-P., Chen, W.-W., Liang, Y.-M., & Li, C.-J. (2009). *Organic Letters, 11*, 3246–3249.
64. Dai, J.-J., Liu, J.-H., Luo, D.-F., & Liu, L. (2011). *Chemical Communications, 47*, 677–679.
65. Feng, C., & Loh, T-P. (2010). *Chemical Communications, 46*, 4779–4781.
66. Li, M.-Z., Wang, C., & Ge, H.-B. (2011). *Organic Letters, 13*, 2062–2064.
67. Li M.-Z., Wang, C., Fang, P., & Ge H.-B. (2011). *Chemical Communications, 47*, 6587–6589.
68. Cornella, J., Lahlali, H., & Larrosa, I. (2010). *Chemical Communications, 46*, 8276–8278.
69. Park, J., Park, E., Kim, A., Park, S-A., Lee, Y., Chi, K-W., Jung, Y. H., & Kim, I. S. (2011). *The Journal of Organic Chemistry, 76*, 2214–2219.
70. Jia, W., & Jiao, N. (2010). *Organic Letters, 12*, 2000–2003.
71. Duan, Z.-Y., Ranjit, S., Zhang, P.-F., & Liu, X.-G. (2009). *Chemistry - A European Journal, 15*, 3666–3669.
72. Ranjit, S., Duan, Z., Zhang, P., & Liu, X. (2010). *Organic Letters, 12*, 4134–4136.
73. Hu, J., Zhao, N., Yang, B., Wang, G., Guo, L.-N., Liang, Y.-M., & Yang, S.-D. (2011). *Chemistry - A European Journal, 17*, 5516–5521.
74. Luo, Y., Pan, X.-L., & Wu, J. (2010). *Tetrahedron Letters, 51*, 6646–6648.
75. Cornella, J., Rosillo-Lopez, M., & Larrosa, I. (2011). *Advanced Synthesis & Catalysis, 353*, 1359–1366.
76. Luo, Y., & Wu, J. (2010). *Chemical Communications, 46*, 3785–3787.

77. Lalic, G., Aloise, A. D., & Shair, M. D. (2003). *Journal of the American Chemical Society*, 125, 2852–2853.
78. Magdziak, D., Lalic, G., Lee, H. M., Fortner, K. C., Aloise, A. D., & Shair, M. D. (2005). *Journal of the American Chemical Society*, *127*, 7284–7285.
79. Yin, L., Kanai, M., & Shibasaki, M. (2009). *Journal of the American Chemical Society*, *131*, 9610–9611.
80. Yamashita, M., Hirano, K., Satoh, T., & Miura, M. (2011). *Advanced Synthesis & Catalysis*, *353*, 631–636.
81. Lindh, J., Sjöberg, P. J. R., & Larhed, M. (2010). Angewandte Chemie International Edition, *49*, 7733–7737.
82. Sun, Z.-M., & Zhao, P.-J. (2009). *Angewandte Chemie International Edition, 48*, 6726–6730.
83. Weaver, J. D., Recio, III A., Grenning, A. J., & Tunge, J. A. (2011). *Chemical Reviews, 111*, 1846–1913.
84. Torregrosa, R. R. P., Ariyarathna, Y., Chattopadhyay, K., Tunge, J. A. (2010). *Journal of the American Chemical Society*, 132, 9280–9282.
85. Shintani, R., Park, S., Shirozu, F., Murakami, M., & Hayashi, T. (2008). *Journal of the American Chemical Society*, *130*, 16174–16175.
86. Goossen, L. J., & Melzer, B. (2007). *The Journal of Organic Chemistry*, *72*, 7473–7476.
87. Goossen, L. J., & Knauber, T. (2008). *The Journal of Organic Chemistry*, *73*, 8631–8634.
88. Shang, R., Ji, D.-S., Chu, L., Fu, Y., & Liu, L. (2011). *Angewandte Chemie International Edition*, *50*, 4470–4474.

Chapter 2
Palladium-Catalyzed Decarboxylative Coupling of Potassium Oxalate Monoester with Aryl and Alkenyl Halides

Abstract Pd-catalyzed decarboxylative cross-couplings of aryl iodides, bromides, and chlorides with potassium oxalate monoesters have been discovered. This reaction is potentially useful for laboratory-scale synthesis of aryl and alkenyl esters. Pd catalyst with bidentate phosphine ligands was found as the optimal catalyst, and unlike other reported decarboxylative couplings, copper is not needed as support catalyst in this reaction. The theoretical calculation shows that the decarboxylation on Pd(II) is the rate-determining step, with a transition state where Pd(II) has a five-coordination state. The calculated energy barrier of rate-determining step is about ~ 30 kcal/mol, which is in accordance with the optimized reaction temperature.

2.1 Introduction

Aromatic esters are important structural elements and synthetic intermediates. Transition metal-catalyzed synthesis of aromatic esters from aryl halides has been studied mainly in the frame of Pd-catalyzed carbonylation [1, 2]. The drawback of

© Springer Nature Singapore Pte Ltd. 2017
R. Shang, *New Carbon–Carbon Coupling Reactions Based on Decarboxylation and Iron-Catalyzed C–H Activation*, Springer Theses,
DOI 10.1007/978-981-10-3193-9_2

handling toxic CO gas and, under many circumstances, the requirement for high pressure reaction conditions often limit the scope of this reaction, especially on a laboratory scale [3]. With an inspiration of decarboxylative coupling, we noticed that we might use some carboxylate salts as d-synthon equivalent, of which the corresponding organometallics are unstable and difficult to prepare. At this juncture, we noticed that potassium oxalate monoesters may be used as an ester synthetic equivalent if this is easily accessible and stable salt can be smoothly decarboxylated in the presence of a metal catalyst. Here we report a novel, practical synthesis of aromatic esters via Pd-catalyzed decarboxylative coupling of oxalate monoester salts with aryl halides (Scheme 2.1).

This study was inspired by the recent seminal work of Goossen and coworkers [3], who discovered the decarboxylative cross-couplings of α-oxocarboxylates giving rise to ketones. Related elegant studies on decarboxylative cross-coupling reactions of aromatic carboxylates have also been reported recently by Myers [4], Forgione [5], Goossen [6], and several other groups [7–11].

2.2 Results and Discussion

2.2.1 Investigation of the Reaction Conditions

Our investigation started by examining the coupling between potassium 2-ethoxy-2-oxoacetate and bromobenzene (Table 2.1). A series of palladium salts and phosphine ligands were examined [12]. When using palladium trifluoroacetate with dppp as the catalyst and phosphine ligands, we got the best results of the reaction. Under the optimal conditions [1 mol% Pd(TFA)$_2$, 1.5 mol% dppp], the desired product was obtained in 85% yield. Not only aryl bromide, aryl iodide as the substrate can also have good results (entry 20). When aryl iodide was used, we found that the desired product can be obtained in 85% yield in the absence of a phosphine ligand, but for aryl chlorides (entry 21), only trace amounts of the desired product can be obtained. It should be mentioned that potassium 2-ethoxy-2-oxoacetate is a stable, crystalline salt, which can be readily made from diethyl oxalate, KOAc, and H$_2$O [13]. Thus, the present protocol is operationally simpler than the previous Pd-catalyzed carbonylation method (usually conducted at 100–150 °C with 1 mol % Pd catalyst [1]) because it avoids the use of toxic CO. This feature is advantageous especially for laboratory-scale synthesis.

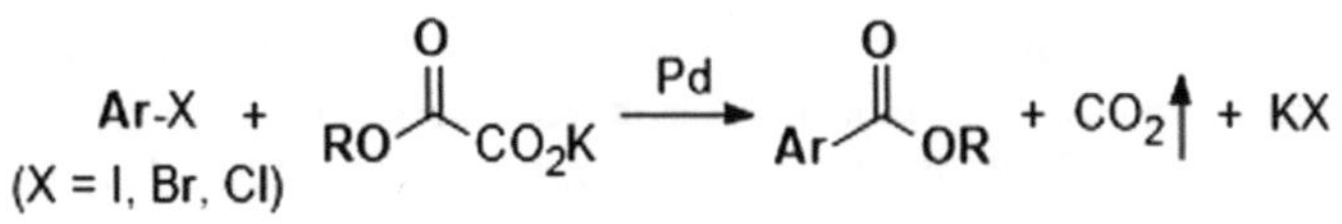

Scheme 2.1 Decarboxylative coupling of potassium oxalate monoesters with aryl halides

Table 2.1 Decarboxylative coupling under various conditions[a]

$$EtO \overset{O}{\underset{}{\parallel}} CO_2K \ + \ Ph-X \xrightarrow[150\ ^\circ C,\ 24\ h,\ NMP]{Pd\ source/ligand} Ph \overset{O}{\underset{}{\parallel}} OEt$$

Entry	X	Pd source	Ligand	Yield %[c]
1	Br	Pd(OAc)$_2$	–	<5
2	Br	Pd(OAc)$_2$	PPh$_3$	27
3	Br	Pd(OAc)$_2$	P(o-Tol)$_3$	51
4	Br	Pd(OAc)$_2$	P(o-MeOPh)$_3$	62
5	Br	Pd(OAc)$_2$	PCy$_3$	9
6	Br	Pd(OAc)$_2$	dppf	23
7	Br	Pd(OAc)$_2$	dppp	81
8	Br	Pd(OAc)$_2$	dppe	68
9	Br	Pd(OAc)$_2$	dppm	21
10	Br	Pd(OAc)$_2$	S-BINAP	36
11	Br	Pd(OAc)$_2$	X-Phos	61
12	Br	Pd(OAc)$_2$	JohnPhos	56
13	Br	PdCl$_2$	dppp	80
14	**Br**	**Pd(TFA)$_2$**	**dppp**	**85**
15	Br	Pd(acac)$_2$	dppp	75
16	Br	Pd$_2$(dba)$_3^b$	dppp	81
17	Br	Pd(dppf)Cl$_2$	dppp	79
18	Br	Pd(PPh$_3$)$_2$Cl$_2$	dppp	80
19	Br	Pd(PPh$_3$)$_4$	dppp	77
20	I	Pd(TFA)$_2$	dppp	83
21	Cl	Pd(TFA)$_2$	dppp	<5

[a]*Conditions* 1 mol% Pd, 3 mol% monodentate ligand or 1.5 mol% bidentate ligand, aryl halide/potassium 2-ethoxy-2-oxoacetate = 1:1.5, 1.0 ml of *N*-methylpyrrolidone (NMP) solvent. All of the reactions were carried out at 0.5 mmol scale. [b]0.5 mol%. [c]GC yields based on PhX *bold* Optimal conditions

The reaction is distinct from reported palladium-catalyzed decarboxylative coupling reactions, such as decarboxylative alkenylation reaction reported by Myers et al. [4] in which oxidative addition of palladium to aryl halide is not involved. It is worth mentioning that in Myers' reaction, addition of phosphine as ligand is detrimental while phosphine ligands play a crucial role in this decarboxylative ester synthesis. Compared with the decarboxylative biaryl synthesis reported by Goossen et al. [6], which was catalyzed by a catalyst composed of copper catalyst for decarboxylation and palladium catalyst for coupling, only a palladium catalyst is used in this reaction, and the Cu catalyst is not necessary, which means both the decarboxylation and coupling took place on palladium.

2.2.2 Exploration of the Substrate Scope

After optimizing the catalyst system, we tested the generality of the reaction with regard to both coupling partners (Table 2.2). It was found that both electron-rich and electron-poor aryl bromides can be successfully converted across a range of

Table 2.2 Decarboxylative cross-coupling with diverse aryl bromides[a]

(reaction scheme):

RO–CO(O)–CO$_2$K (R = Me, Et) 1.1~1.5 mmol[b] + Ar–Br 1.0 mmol → 1 mol% Pd(TFA)$_2$, 1.5 mol% dppp, NMP (2 mL), 150 °C, 16~24 h → RO–C(O)–Ar

dppp = Ph$_2$P–CH$_2$CH$_2$CH$_2$–PPh$_2$

Compound	Yield
1	83%
2	96%
3	92%
4	95%
5	79%
6	90%
7	89%
8	75%
9	82%
10	64%
11	67%
12	78%
13	86%
14	98%
15	52%
16	80%
17	94%
18	80%
19	82%
20	81%
21	62%
22	94%
23	81%
24	54%
25	55%
26	70%
27	77%
28	81%

[a]Isolated yields based on aryl bromides. [b]See the supporting information

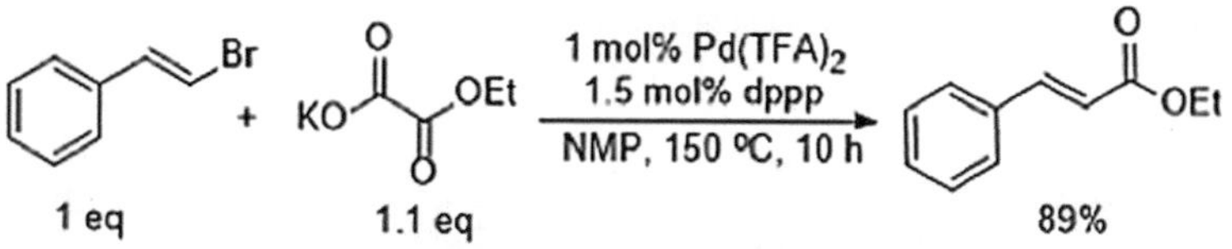

Scheme 2.2 Synthesis of *trans*-acrylate derivatives

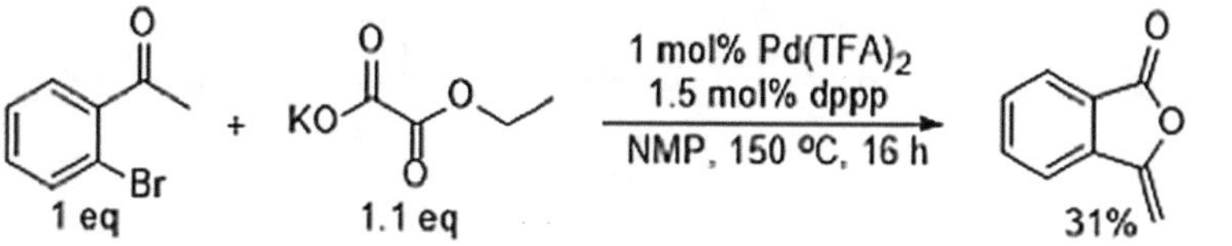

Scheme 2.3 Casade cross-coupling/cyclization

functional groups [including ether (entries 4, 5, 7), thioether (entry 8), aldehyde (entry 11), ketone (entries 10, 14), amide (entry 15), nitro (entry 9), nitrile (entry 13), ester (entry 12), trifluoromethyl (entries 16, 17, 18), halogen (entries 22, 23) and heterocycle (entry 19)]. Importantly, ortho substitution can be tolerated in the transformation (entries 3, 5, 15). In addition to the ethyl esters, potassium 2-methoxy-2-oxoacetate can be used to produce methyl esters (entries 24–28). Furthermore, the method can be used to synthesize *trans*-acrylate derivatives in high yields from vinyl bromides (Scheme 2.2), and in a special case (Scheme 2.3), we observed cascade cross-coupling/cyclization.

The above protocol can be applied to both aryl bromides and iodides (Table 2.1, entry 20) but not to aryl chlorides (entry 21). Use of bulky, electron-rich ligands may solve the problem, but our experiments with good Ar–Cl activation ligands such as t-Bu$_3$P, S-Phos, DavePhos, X-Phos, and JohnPhos [14] failed to couple PhCl with potassium 2-ethoxy-2-oxoacetate. We reasoned that the use of a bulky, electron-rich ligand similar in structure to dppp might provide a solution. To our delight, dCypp proved to successfully promote the decarboxylative coupling with various aryl chlorides (Table 2.3). However, *ortho* substitution would inhibit the reaction due to the bulky steric hindrance of the dCypp (Table 2.3, entry 3).

2.2.3   Mechanistic Study

Standard density functional theory methods were used to understand the mechanism of the new decarboxylative cross-coupling reaction (Fig. 2.1) [15]. First, a Pd(0) complex was proposed to activate the aryl halide. When 1,3-diphosphinopropane was used as a model ligand, the Pd(0) complex formed a η^2 complex with PhBr (**IN1**), which should undergo oxidative addition through **TS1** to produce a

Table 2.3 Decarboxylative cross-coupling with aryl chlorides[a]

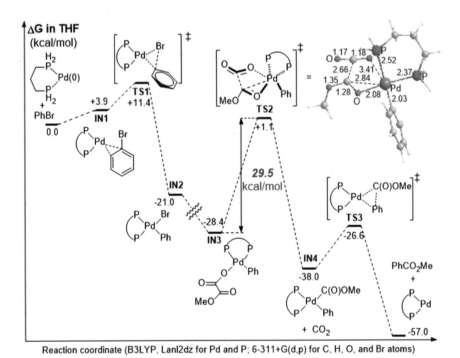

[a]Isolated yields based on aryl chlorides

Fig. 2.1 Proposed mechanism of decarboxylative cross-coupling. Reprinted with the permission from J. Am. Chem. Soc. 2009, 131, 5738. Copyright 2011 American Chemical Society

four-coordinate Pd(II) intermediate (**IN2**). The energy barrier for oxidative addition was +11.4 kcal/mol. **IN2** then exchanged the anion to form **IN3**. From **IN3**, the decarboxylation transition state (**TS2**) was indentified as a five-coordinate Pd(II) species [16]. In **TS2**, the Pd(II) coordinated to the leaving CO_2 moiety through one of its oxygens and to the other carbonyl group in an η^2 mode. From **IN3** to **TS2**, the free energy increased by 29.5 kcal/mol, a value whose magnitude is consistent with the experimental temperature required for the reaction (~ 150 °C). Thus, decarboxylation is the rate-limiting step in the catalytic cycle. The immediate product of decarboxylation was a four-coordinate acyl-Pd complex (**IN4**), which readily underwent reductive elimination to produce the ester product through **TS3** with a low barrier of ~ 12 kcal/mol. In **IN4**, the Pd center is coordinatively saturated, preventing decarbonxylation to form an inactive Pd–CO complex. This may explain why bidentate phosphine ligands are favored for the present decarboxylative cross-coupling reactions. A related phenomenon has been discussed for Pd-catalyzed carbonylation [17], where the use of bulky, electron-rich bidentate ligands has also been found to be important [2].

2.3 Conclusion

Pd-catalyzed decarboxylative cross-coupling of aryl iodides, bromides, and chlorides with potassium oxalate monoesters has been discovered. This reaction is potentially useful for laboratory-scale synthesis of aryl and alkenyl esters [18]. Bulky, electron-rich bidentate phosphine ligands are preferred in the reaction, whereas Cu is not needed for decarboxylation. Theoretical calculations suggest a five-coordinate Pd(II) transition state for decarboxylation with an energy barrier of ~ 30 kcal/mol.

2.4 Experimental Section and Compound Data

2.4.1 General Information

All reactions were carried out in oven-dried Schlenk tubes under Argon atmosphere (purity $\geq 99.999\%$). The NMP solvent was bought from Alfa Aesar (sealed under argon) without further purification. All aryl halides were purchased from Alfa Aesar or Acros and used directly. All phosphine ligands were bought from Sigma-Aldrich, Strem, or Alfa Aesar and sealed under Argon. All the other reagents and solvents were bought from Sinopharm Chemical Reagent Co. Ltd or Alfa Aesar and were purified when necessary.

^{1}H-NMR, ^{13}C-NMR spectra were recorded on a Bruker Advance 400 spectrometer at ambient temperature in $CDCl_3$ unless otherwise noted. Data for

^{1}H-NMR are reported as follows: chemical shift (δ ppm), multiplicity, integration, and coupling constant (Hz). Data for ^{13}C-NMR are reported in terms of chemical shift (δ ppm). Gas chromatographic (GC) analysis was acquired on a Shimadzu GC-2014 Series GC System equipped with a flame ionization detector. GC-MS analysis was performed on Thermo Scientific AS 3000 Series GC-MS System. MS analysis was performed on Finnigan LCQ advantage Max Series MS System. Elementary Analysis was carried out on Elementar Vario EL III elemental analyzer. Organic solutions were concentrated under reduced pressure on a Buchi rotary evaporator. Flash column chromatographic purification of products was accomplished using forced-flow chromatography on Silica Gel (200–300 mesh).

2.4.2 Experimental Procedure

General procedure A: the synthesis of aromatic esters from aryl bromides

Palladium(II) trifluoroacetate (0.01 mmol), 1,3-bis(diphenylphosphino)-propane (0.015 mmol), appointed amount of ethyl (or methyl) potassium oxalate (1.1–1.5 mmol, see Page S8-16 for detailed quantities) and the aryl bromide (1.00 mmol) (if solid) were placed in an oven-dried 20 ml Schlenk tube. The reaction vessel was evacuated and filled with argon for three times. Then aryl bromide (1.00 mmol) (if liquid) and NMP (2 ml) were added with a syringe under a counter flow of argon. The vessel was sealed, connected to the Schlenk line which was full with argon and stirred at 150 ± 5 °C for the appointed time. Upon completion of the reaction, the mixture was cooled to room temperature and diluted with diethyl ether (20 ml). It was then filtered through a short silica column to remove the deposition. The organic layers were washed with water (20 ml $\times$ 3), and then with brine, dried over Na_2SO_4, and filtered. The solvents were removed. Purification of the residue by column chromatography (silica gel, ethyl acetate/hexane gradient) yielded the corresponding aryl ester.

General procedure B: the synthesis of aromatic esters from aryl chlorides

Palladium(II) trifluoroacetate (0.03 mmol), appointed amount of ethyl potassium oxalate and the aryl chloride (1.00 mmol) (if solid) were placed in an oven-dried 20 ml Schlenk tube. The reaction vessel was evacuated and filled with argon for three times. Aryl chloride (1.00 mmol) (if liquid), 1,3-bis(dicyclohexylphosphino) propane (0.06 mmol, 108 μl[1]) (as a solution, 250 mg in 2 ml NMP) and NMP (2 ml) were added with a syringe under a counter flow of argon, the vessel was sealed, connected to the Schlenk line which was full with argon and stirred at 150 ± 5 °C for the appointed time. Upon completion of the reaction, the mixture was cooled to room temperature, diluted with diethyl ether (20 ml) and filtered

---

[1]1,3-bis(dicyclohexylphosphino)propane was bought from Sigma-Aldrich as a colorless oil (250 mg Package), and 2 ml NMP was directly added to the bottle to dissolve it before use.

through a short silica column to remove the deposition. The organic layers were washed with water (20 ml × 3) and then with brine, dried over Na_2SO_4, and filtered. The solvents were removed. Purification of the residue by column chromatography (silica gel, ethyl acetate/hexane gradient) yielded the corresponding aryl ester.

2.4.3 Characterization of the Products

Compound name benzoic acid ethyl ester

colorless liquid (125 mg, 83% yield). 1H-NMR (400 MHz, CDCl3): δ 1.40 (t, 3H, $J = 7.2$ Hz), 4.38 (q, 2H, $J = 7.1$ Hz), 7.41–7.46 (m, 2H), 7.52–7.57 (m, 1H), 8.03–8.06 (m, 2H). 13C-NMR (100 MHz, CDCl3, δ ppm): 14.3, 60.9, 128.3, 129.5, 130.6, 132.8, 166.6.

Compound name Benzo [1,3] dioxole-5-carboxylic acid ethyl ester

yellow liquid (173 mg, 89% yield). Spectral data matched literature description (Ref. Lee, A. S.; Wu, C.C.; Lin, L.S.; Hsu, H.F. Synthesis. 2004, 4, 568). 1H-NMR (400 MHz, CDCl3): δ 1.37 (t, 3H, $J = 7.0$ Hz), 4.33 (q, 2H, $J = 7.2$ Hz), 6.02 (s, 2H), 6.82 (d, 1H, $J = 8.4$ Hz), 7.46 (d, 1H, $J = 1.6$ Hz), 7.65 (dd, 1H, $J1 = 1.6$ Hz, $J2 = 8.0$ Hz). 13C-NMR (100 MHz, CDCl3, δ ppm): 14.4, 60.9, 101.8, 107.9, 109.5, 124.6, 125.3, 147.7, 151.5, 166.0.

Compound name 4-methylthiobenzoic acid ethyl ester

yellow acicular solid (147 mg, 75% yield). Spectral data matched literature description (Ref. Lu, X.; Rodriguez, M.; Gu, W.; Silverman, R. B. Bioorg. Med. Chem. 2003, 11, 4423). 1H-NMR (400 MHz, CDCl3): δ 1.38 (t, 3H, $J = 7.2$ Hz), 2.51 (s, 3H), 4.35 (q, 2H, $J = 7.1$ Hz), 7.24 (d, 2H, $J = 8.40$ Hz), 7.93 (d, 2H, $J = 8.8$ Hz). 13C-NMR (100 MHz, CDCl3, δ ppm): 14.4, 14.9, 60.9, 125.0, 126.7, 129.9, 145.3, 166.6.

Compound name 4-(trifluoromethyl)benzoic acid ethyl ester

colorless liquid (175 mg, 80% yield). Spectral data matched literature description (Ref. Bromilow, J.; Brownlee, R.; Craik, D. J.; Sadek, M.; Taft, R. W. J. Org. Chem. 1980, 45, 2429). 1H-NMR (400 MHz, CDCl3): δ 1.42 (t, 3H, J = 7.2 Hz), 4.41 (q, 2H, J = 7.1 Hz), 7.69 (d, 2H, J = 8.4 Hz), 8.15 (dd, 2H, $J1$ = 0.4 Hz, $J2$ = 8.8 Hz). 13C-NMR (100 MHz, CDCl3, δ ppm): 14.3, 61.6, 123.8 (q, J = 271 Hz), 125.4 (q, J = 3.5 Hz), 130.0, 133.8, 134.4 (q, J = 32.4 Hz), 165.4.

Compound name 4-trifluoromethoxy-benzoic acid ethyl ester

colorless liquid (187 mg, 80% yield). 1H-NMR (400 MHz, CDCl3): δ 1.40 (t, 3H, J = 7.2 Hz), 4.39 (q, 2H, J = 7.1 Hz), 7.26 (d, 2H, J = 8.8 Hz), 8.09 (d, 2H, J = 8.4 Hz). 13C-NMR (100 MHz, CDCl3, δ ppm): 14.3, 61.4, 120.4 (q, J = 257), 120.3, 129.0, 131.6, 152.6, 165.5.

Compound name 3-pyridinecarboxylic acid ethyl ester

colorless liquid (124 mg, 82% yield). 1H-NMR (400 MHz, CDCl3): δ 1.42 (t, 3H, J = 7.0 Hz), 4.42 (q, 2H, J = 7.1 Hz), 7.39 (m, 1H), 8.31 (m, 1H), 8.78 (m, 1H), 9.23 (d, 1H, J = 2.0 Hz). 13C-NMR (100 MHz, CDCl3, δ ppm): 14.2, 61.3, 123.2, 126.3, 136.9, 150.8, 153.2, 165.1.

Compound name naphthalene-1-carboxylic acid ethyl ester

yellow liquid (124 mg, 62% yield). 1H-NMR (400 MHz, CDCl3): δ 1.44 (t, 3H, J = 7.2 Hz), 4.46 (q, 2H, J = 7.2 Hz), 7.44–7.52 (m, 2H), 7.57–7,61 (m, 1H), 7.83–7.87 (d, 1H, J = 8.4 Hz), 7.96–7.98 (d, 1H, J = 8.4 Hz), 8.16–8.18 (dd, 1H, $J1$ = 7.2 Hz, $J2$ = 1.2 Hz), 8.91–8.93 (d, 1H, J = 8.4 Hz). 13C-NMR (100 MHz, CDCl3, δ ppm): 14.4, 61.1, 124.5, 125.9, 126.2, 127.6, 127.7, 128.6, 130.1, 131.4, 133.2, 133.9, 167.6.

Compound name 4-acetylbenzoic acid ethyl ester

white solid (123 mg, 64% yield). Spectral data matched literature description (Ref. Cai, C.; Rivera, N. R.; Balsells, J.; Sidler, R. R.; McWilliams, J. C.; Shultz, C. S.; Sun, Y. Org. Lett. 2006, 8, 5161). 1H-NMR (400 MHz, CDCl3): δ 1.42 (t, 3H, J = 7.0 Hz), 2.64 (s, 3H), 4.41 (q, 2H, J = 7.0 Hz), 8.01 (m, 2H), 8.12 (m, 2H). 13C-NMR (100 MHz, CDCl3, δ ppm): 14.4, 26.9, 61.5, 128.2, 129.8, 134.4, 140.2, 165.8, 197.6.

Compound name 4-cyanobenzoic acid ethyl ester

yellow solid (150 mg, 86% yield). 1H-NMR (400 MHz, CDCl3): δ 1.42 (t, 3H, J = 7.2 Hz), 4.42 (q, 2H, J = 7.2 Hz), 7.74 (d, 2H, J = 8.0 Hz), 8.14 (d, 2H, J = 8.0 Hz). 13C-NMR (100 MHz, CDCl3, δ ppm): 14.3, 61.9, 116.4, 118.0, 130.1, 132.2, 134.4, 165.0.

Compound name 4-ethoxycarbonylbenzophenone

yellow oil (249 mg, 98% yield). Spectral data matched literature description (Ref. Duplais, C.; Bures, F.; Sapountzis, I.; Korn, T. J.; Cahiez, G.; Knochel, P. Angew. Chem. 2004, 116, 3028; Angew. Chem. Int. Ed. 2004, 43, 2968). 1H-NMR (400 MHz, CDCl3): δ 1.42 (t, 3H, J = 7.2 Hz), 4.42 (q, 2H, J = 7.2 Hz), 7.48–7.52 (m, 2H), 7.59–7.64 (m, 1H), 7.79–7.85 (m, 4H), 8.14–8.17 (m, 2H). 13C-NMR (100 MHz, CDCl3, δ ppm): 14.4, 61.5, 128.5, 129.5, 129.8, 130.2, 133.0, 133.7, 137.1, 141.3, 165.9, 196.1.

Compound name 3-methylidene-3H-isobenzofuran-1-one

white solid (45 mg, 31% yield). Spectral data matched literature description (Ref. Yamamoto, H.; Pandey, G.; Asai, Y.; Nakano, M.; Kinoshita, A.; Namba, K.; Imagawa, H.; Nishizawa, M. Org. Lett. 2007, 9, 4029) 1H-NMR (400 MHz, CDCl3): δ 5.23 (dd, 2H, $J1$ = 3.2 Hz, $J2$ = 4.0), 7.55–7.62 (m, 1H), 7.72–7.73 (m,

2H), 7.90–7.92 (m, 1H). 13C-NMR (100 MHz, CDCl3, δ ppm): 91.3, 120.7, 125.2, 125.4, 130.5, 134.5, 139.1, 151.9, 166.9.

Compound name 2-acetylamino benzoic acid ethyl ester

pale yellow solid (108 mg, 52% yield). 1H-NMR (400 MHz, CDCl3): δ 1.42 (t, 3H, J = 7.0 Hz), 2.24 (s, 3H), 4.38 (q, 2H, J = 7.2 Hz), 7.05–7.09 (m, 1H), 7.51–7.56 (m, 1H), 8.03–8.05 (m, 1H), 8.69–8.71 (m, 1H), 11.09 (s, br, 1H). 13C-NMR (100 MHz, CDCl3, δ ppm): 14.3, 25.6, 61.5, 115.2, 120.4, 122.4, 130.8, 134.6, 141.7, 168.4, 169.1.

References

1. A recent review: Barnard, C. F. J. (2008). *Organometallics, 27*, 5402–5422.
2. Munday, R., Martinelli, J., & Buchwald, S. L. (2008). *Journal of the American Chemical Society, 130*, 2754–2755.
3. Odell, L. R., Saevmarker, J., & Mats, L. (2008). *Tetrahedron Letters, 49*, 6115–6118.
4. (a) Myers, A. G., Tanaka, D., & Mannion, M. (2002). *Journal of the American Chemical Society, 124*, 11250–11251. (b) Tanaka, D., Romeril, S., & Myers, A. G. (2005). *Journal of the American Chemical Society, 127*, 10323–10333.
5. Forgione, P., Brochu, M. C., St-Onge, M., Thesen, K. H., Bailey, M. D., & Bilodeau, F. (2006). *Journal of the American Chemical Society, 128*, 11350–11351.
6. (a) Goossen, L. J., Deng, G., & Levy, L. M. (2006). *Science, 313*, 662–664. (b) Goossen, L. J., Rodriguez, N., Melzer, B., Linder, C., Deng, G., & Levy, L. M. (2007). *Journal of the American Chemical Society, 129*, 4824–4833. (c) Goossen, L. J., Rodriguez, N., & Linder, C. (2008). *Journal of the American Chemical Society, 130*, 15248–15249.
7. (a) Becht, J.-M., Catala, C., Le Drian, C., & Wagner, A. (2007). *Organic Letters, 9*, 1781–1783. (b) Becht, J. M., & Le Drian, C. (2008). *Organic Letters, 10*, 3161–3164.
8. Dickstein, J. S., Mulrooney, C. A., O'Brien, E. M., Morgan, B. J., & Kozlowski, M. C. (2007). *Organic Letters, 9*, 2441–2444.
9. (a) Moon, J., Jeong, M., Nam, H., Ju, J., Moon, J. H., Jung, H. M., & Lee, S. (2008). *Organic Letters, 10*, 945–948. (b) Maehara, A., Tsurugi, H., Satoh, T., & Miura, M. (2008). *Organic Letters, 10*, 1159–1162.
10. Voutchkova, A., Coplin, A., Leadbeater, N. E., & Crabtree, R. H. (2008). *Chemical Communications*, 6312–6314.
11. For related studies see: (a) Weaver, J. D., & Tunge, J. A. (2008). *Organic Letters, 10*, 4657–4660. (b) Waetzig, S. R., & Tunge, J. A. (2008). *Chemical Communications*, 3311–3313. (c) Burger, E. C., & Tunge, J. A. (2006). *Journal of the American Chemical Society, 128*, 10002–10003. (d) Rayabarapu, D. K., & Tunge, J. A. (2005). *Journal of the American Chemical Society, 127*, 13510–13511.
12. Cu co-catalyst was used in the beginning but our experiments quickly revealed that the use of Pd only is sufficient for the present system.

13. Klemm, L. H., & Lu, J. J. (1986). *Organic Preparations and Procedures International, 18*, 237–244.
14. Surry, D. S., & Buchwald, S. L. (2008). *Angewandte Chemie International Edition, 47*, 6338–6361.
15. Goossen, L., Thiel, W. R., Rodriguez, N., Linder, C., & Melzer, B. (2007). *Advanced Synthesis & Catalysis, 349*, 2241–2246.
16. Stromnova, T. A., & Moiseev, I. I. (1998). *Russian Chemical Reviews, 67*, 485–514.
17. Hama, T., Liu, X., Culkin, D. A., & Hartwig, J. F. (2003). *Journal of the American Chemical Society, 125*, 11176–11177.

Chapter 3
Synthesis of Polyfluorobiaryls via Copper-Catalyzed Decarboxylative Couplings of Potassium Polyfluorobenzoates with Aryl Bromides and Iodides

Abstract In this chapter, we report a copper-catalyzed decarboxylative cross-coupling of potassium polyfluorobenzoates with aryl iodides and bromides. This reaction can be used for the preparation of polyfluoroaromatic compounds and polyfluorostilbene. The reactants used in the reaction are easily accessible non-volatile polyfluorobenzoate salts. Mechanistic studies suggest that decarboxylation occurs at first on copper(I) to generate a polyfluorophenylcopper(I) intermediate, which then undergoes oxidative addition with aryl halides and reductive elimination to produce the coupling products.

3.1 Introduction

Transition metal-catalyzed decarboxylative cross-coupling using carboxylic acids as aryl sources is of contemporary interest [1]. This method avoids the use of expensive and sensitive organometallic reagents, and generates CO_2 instead of toxic metal halides. Early studies showed that a stoichiometric quantity of copper could promote the decarboxylative coupling of aromatic carboxylic acids with aryl iodides [2]. Until recently, Goossen et al. reported Pd/Cu-catalyzed decarboxylative coupling of benzoic acids and α-oxocarboxylates with aryl halides and triflates [3]. Related studies by the groups of Myers [4], Forgione [5], and others [6–9] showed that

© Springer Nature Singapore Pte Ltd. 2017
R. Shang, *New Carbon–Carbon Coupling Reactions Based on Decarboxylation and Iron-Catalyzed C–H Activation*, Springer Theses,
DOI 10.1007/978-981-10-3193-9_3

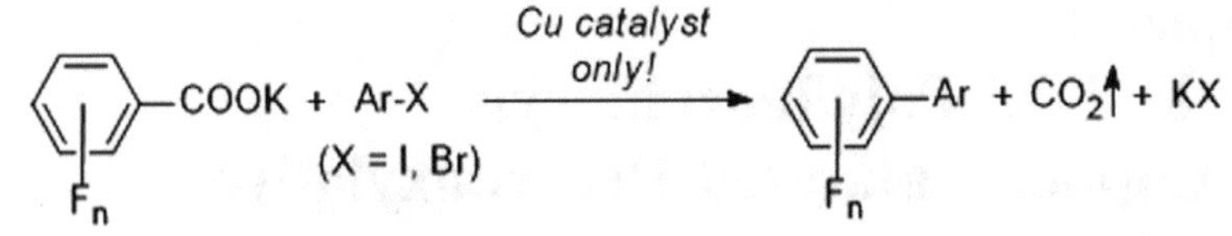

Scheme 3.1 Decarboxylative coupling of potassium polyfluorobenzoate with aryl halides

palladium by itself could also catalyze the decarboxylative coupling of aromatic carboxylic acids. We reported a palladium-catalyzed decarboxylative coupling of oxalate monoesters with aryl halides [10], and related decarboxylative reactions were also reported by the groups of Tunge, Li, Chruma, and others recently [11].

In order to further expand the scope of decarboxylation coupling reaction, herein we describe the first, copper-only systems that catalyze the decarboxylative coupling of potassium polyfluorobenzoates with aryl iodides and bromides (Scheme 3.1) [12]. The importance of the study is twofold: (1) The new reactions can replace the use of expensive but often less reactive [13] fluorobenzene organometallics in the synthesis of polyfluorobiaryls, which are important molecules in medicinal chemistry [14] and material science [15]. They also provide a method complementary to that reported by Fagnou [16] and Daugulis [17] for fluorobiaryl synthesis through C–H arylation of a polyfluoroarene. (2) Recently Goossen et al. reported the copper-catalyzed protodecarboxylation of aromatic carboxylic acids [18]. Nonetheless, there has been no example for copper-catalyzed decarboxylative coupling of acids with aryl halides. Therefore, the reactions reported herein represent a novel type of copper-catalyzed cross-coupling reactions [19, 20].

3.2   Results and Discussion

3.2.1   Investigation of Reaction Condition

Our work started with the decarboxylative coupling of C_6F_5COOK with PhI. When 10 mol% of CuI/1,10-phenanthroline was used as the catalyst, decarboxylation could proceed rapidly in NMP at 160 °C, we could observe a lot of bubbles from the solution, but unfortunately the yield of the desired product was only around 10%. We suspected this might be caused by the too fast decarboxylation rate to match the rate of the coupling reaction at this temperature. To improve the yield, we lowered the reaction temperature and found that the best compromise between the reaction rate and yield is achieved at 130 °C. However, the optimal yield remained at about 40% after we examined many combinations of solvents and ligands. A breakthrough was then made when diglyme was used as the solvent, and the optimal yield obtained was 99% with or even without the ligand. Apparently without the addition of 1,10-phenanthroline as ligand under the reaction conditions, diglyme solvent itself acted as a ligand for copper catalyst. A possible explanation for the outstanding performance of diglyme is that diglyme can coordinate to K^+, thereby facilitating the complexation between CuI and $C_6F_5CO_2^-$.

3.2.2   Exploration of the Substrate Scope

Extending the model reaction to other substrates showed that both electron-rich and electron-poor aryl iodides could be successfully converted and a range of functional groups (such as methoxy, chlorine, nitro, ester, trifluoromethyl, cyano) were tolerated (Table 3.1, entries 2–12). The reaction yields range from good to excellent. Importantly, ortho substitution can be well tolerated in the transformation (Table 3.1, entries 13–16). In addition to phenyl iodides, other aryl substrates including naphthyl, perfluorophenyl, and heteroaryl iodides can also be used to produce the corresponding polyfluorobiaryls (Table 3.1, entries 18–24).

The above protocol can be applied to aryl iodides but not aryl bromides. To solve this problem 1,10-phenanthroline was used as a ligand. As shown in Table 3.2, copper-catalyzed decarboxylative cross-couplings between potassium pentafluorobenzoate and a variety of aryl bromides display high yields ranging from 88 to 95%. These coupling reactions can tolerate both electron-rich and electron-poor substrates (Table 3.2, entries 1–9) and can also tolerate ortho substitution (Table 3.2, entries 14–16). In addition, heteroaryl bromides are suitable substrates in the reaction (Table 3.2, entries 17–20).

Notably, copper-catalyzed decarboxylative cross-coupling of potassium pentafluorobenzoate with 1-bromoadamantane produces a $C(sp^3)$–$C(sp^2)$ bond (Scheme 3.2). Moreover, the method can be used to synthesize polyfluorostilbene in high yields from vinyl bromides (Scheme 3.3).

The scope of the reaction with respect to fluoroarene is presented in Table 3.3. Although diglyme is important for the reactions with potassium pentafluorobenzoate, dimethyl acetamide (DMA) affords better results for fluoroarenes containing fewer fluorine atoms. Under the optimized reaction conditions, potassium monofluorobenzoate cannot be efficiently converted, unless an *ortho*-CF$_3$ group is added (Table 3.3, entries 1–3). Once two F atoms are placed at each of the ortho positions, the decarboxylative coupling of potassium bis(fluorobenzoate) can proceed smoothly with both electron-rich and electron-poor aryl iodides (Table 3.3, entries 4–7). 2,6-difluorobenzoate can also react with 2-methoxy-iodobenzene which has ortho steric hindrance in 89% yield. Similar reactions are also observed with tri- and tetrafluorobenzoates having two ortho-F atoms (Table 3.3, entries 10–18). Such as 2,3,6-trifluorobenzoate, 2,4,6-trifluorobenzoate, 2,3,5,6-tetrafluorobenzoate, and 4-methyl-2,3,5,6-tetrafluorobenzoate can react with a variety of electron-rich and electron-deficient aryl iodides smoothly. A series of functional groups, such as cyano, ester, nitro, trifluoromethyl can successfully be compatible with the reaction. In entries 14 and 15 of Table 3.3 some diarylated by-products are also observed. This means that the direct arylation of acidic C–H bonds of polyfluoroarenes [16, 17] is a side reaction in the copper-catalyzed decarboxylative coupling of polyfluorobenzoates.

Table 3.1 Copper-catalyzed decarboxylative cross-coupling between potassium polyfluorobenzoates and aryl iodides[a]

Entry	Product	Yield%	Entry	Product	Yield%
1	C_6F_5—(phenyl)	99	13[c]	C_6F_5—(o-methyl phenyl)	96
2	C_6F_5—(p-methyl phenyl)	99	14[c]	C_6F_5—(MeO, Cl substituted phenyl)	97
3	C_6F_5—(phenyl)—OMe	99	15	C_6F_5—(phenyl)	94
4	C_6F_5—(phenyl)—Cl	95	16	C_6F_5—(MeO$_2$C substituted phenyl)	99
5	C_6F_5—(phenyl)—NO$_2$	98	17	C_6F_5—(benzodioxane)	99
6	C_6F_5—(phenyl)—CO$_2$Et	99	18[c]	C_6F_5—(naphthyl)	99
7[b]	C_6F_5—(phenyl)—CF$_3$	99	19[b]	C_6F_5—(tetrafluorophenyl)—F	80
8	C_6F_5—(phenyl)—CN	99	20[b]	C_6F_5—(phenyl)—F	94
9	C_6F_5—(OMe substituted phenyl)	96	21	C_6F_5—(pyridyl)	99
10	C_6F_5—(CO$_2$Me substituted phenyl)	98	22	C_6F_5—(pyridyl)	99
11[b]	C_6F_5—(CF$_3$ substituted phenyl)	92	23	C_6F_5—(thienyl)—NO$_2$	61
12	C_6F_5—(NO$_2$ substituted phenyl)	99	24[d]	C_6F_5—(phenyl)—C_6H_5	89

[a]Reactions carried out at 0.5 mmol scale in 0.5 ml diglyme. Isolated yields were calculated from aryl iodides. [b]1.2 eq. $C_6F_5CO_2K$ was used. [c]20 mmol% CuI was used. [d]1,4-di-iodobenzene was used as substrate

3.2.3 Mechanistic Study

Goossen et al. previously conducted a theoretical study on the mechanism of copper-mediated decarboxylation of benzoic acids [21]. For the copper-catalyzed decarboxylative coupling described herein, there is a key mechanistic question as to

Table 3.2 Copper-catalyzed decarboxylative cross-coupling between potassium polyfluorobenzoates and aryl bromides[a]

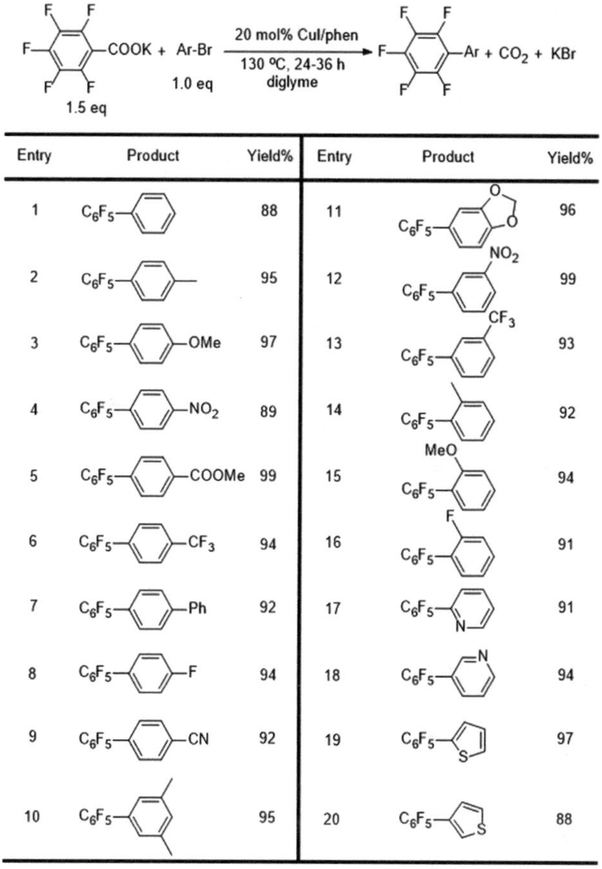

Entry	Product	Yield%	Entry	Product	Yield%
1	C_6F_5—⟨⟩	88	11	C_6F_5—benzodioxole	96
2	C_6F_5—⟨⟩—CH₃	95	12	C_6F_5—⟨⟩-NO₂	99
3	C_6F_5—⟨⟩—OMe	97	13	C_6F_5—⟨⟩-CF₃	93
4	C_6F_5—⟨⟩—NO₂	89	14	C_6F_5—⟨⟩	92
5	C_6F_5—⟨⟩—COOMe	99	15	C_6F_5—⟨⟩-MeO	94
6	C_6F_5—⟨⟩—CF₃	94	16	C_6F_5—⟨⟩-F	91
7	C_6F_5—⟨⟩—Ph	92	17	C_6F_5—⟨⟩N	91
8	C_6F_5—⟨⟩—F	94	18	C_6F_5—⟨⟩=N	94
9	C_6F_5—⟨⟩—CN	92	19	C_6F_5—⟨S⟩	97
10	C_6F_5—⟨⟩	95	20	C_6F_5—⟨S⟩	88

[a]Reactions were carried out at 0.5 mmol scale in 0.5 ml diglyme. Isolated yields were calculated from bromides

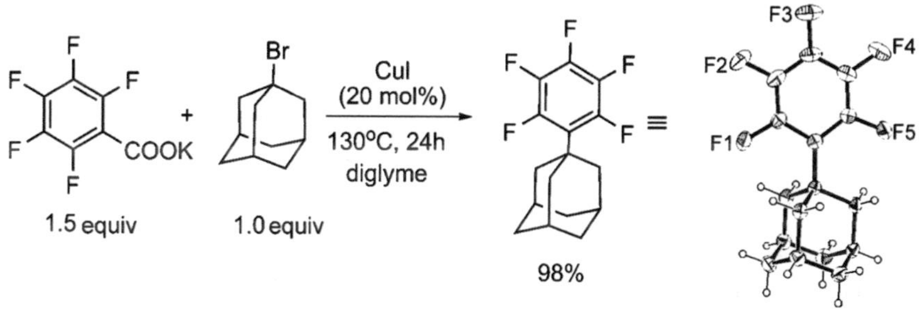

Scheme 3.2 Decarboxylative coupling of potassium pentafluorobenzoate with 1-bromoadamantane

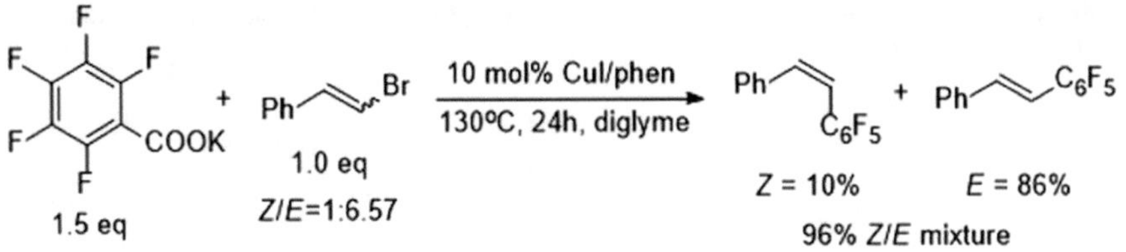

Scheme 3.3 Decarboxylative coupling of potassium pentafluorobenzoate with vinyl bromides

Table 3.3 Copper-catalyzed decarboxylative cross-coupling between aryl iodides and other polyfluorobenzoates[a]

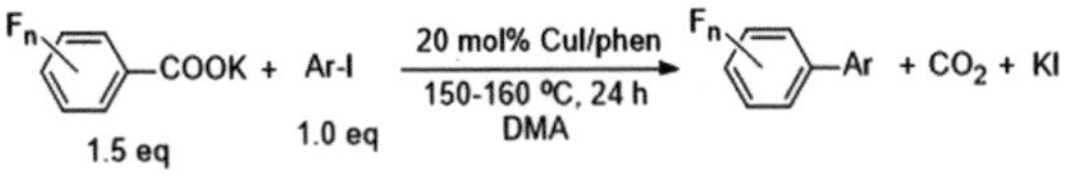

Entry	Product	Yield%	Entry	Product	Yield%
1	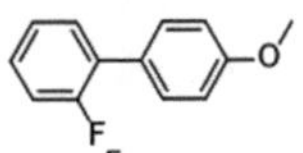	trace	10	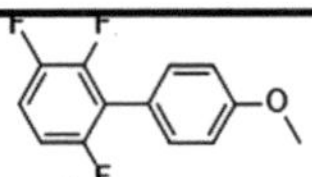	73
2	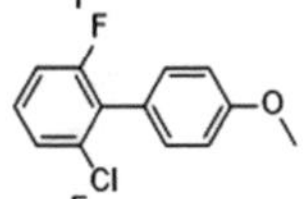	trace	11	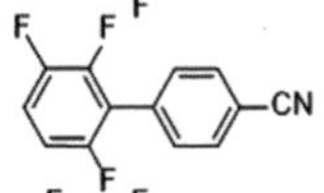	95
3	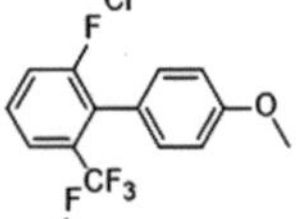	68	12	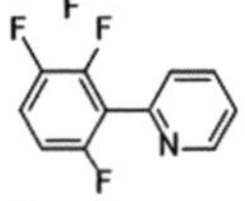	80
4	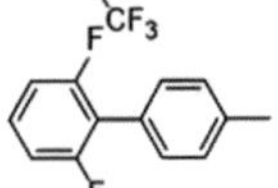	73	13	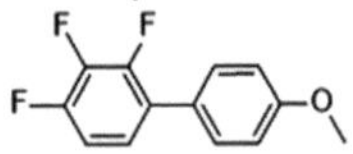	21
5	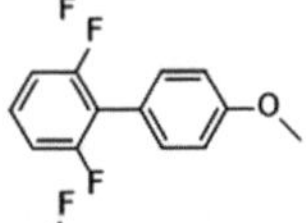	78	14[b]	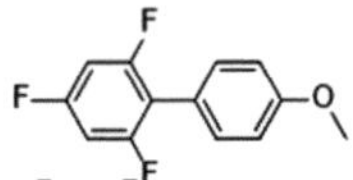	48
6	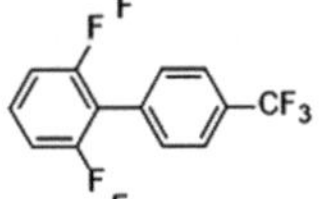	86	15[c]	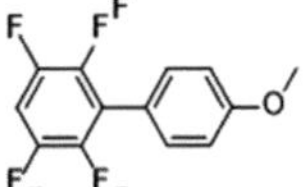	69
7	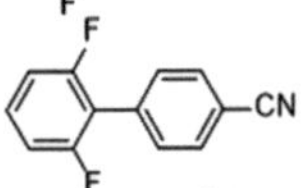	83	16	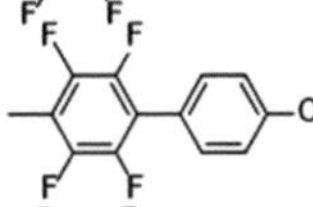	97
8	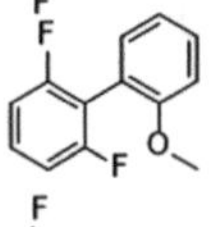	89	17	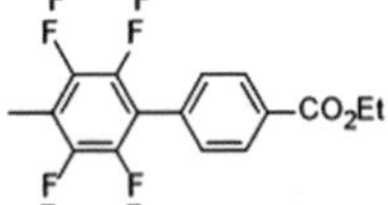	95
9	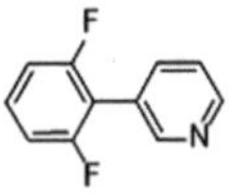	88	18	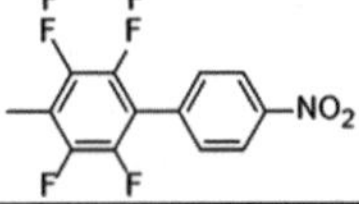	82

[a]Reactions were carried out at 0.25 mmol scale in 0.5 ml DMA. Isolated yields were calculated from aryl iodides. [b]46% of diarylated product was also isolated. [c]29% of diarylated product was also isolated. See supporting information for more details

whether decarboxylation occurs on copper(I) before oxidative addition, or at the copper(III) stage. To solve the problem DFT calculations were performed to compare the two plausible mechanisms (Fig. 3.1) [22]. In pathway I, the initial copper(I) complex **1** reacts with perfluorobenzoate to form **2**, which then undergoes decarboxylation via the four-membered-ring transition state $TS_{(2-3)}$ (Fig. 3.1), which has an energy barrier of +20.3 kcal mol^{-1}. This step produces a new copper(I) complex **3** which can react with PhBr through oxidative addition [23] (transition state $TS_{(3-4)}$). The energy barrier of oxidative addition is +30.0 kcal mol^{-1} and therefore, oxidative addition constitutes the rate-limiting step in pathway I. Finally, reductive elimination is found to be a facile step and it finishes the reaction producing C_6F_5Ph.

In pathway II, oxidative addition takes places first on the complex **1**. This step has a relatively low energy barrier of +18.9 kcal/mol. After oxidative addition the resulting copper(III) species is pentacoordinated and therefore, decarboxylation at copper(III) has to pass through a hexacoordinated transition state. As a result of the strong steric repulsion in the hexacoordinated species, the energy barrier for decarboxylation is calculated to be +51.1 kcal mol^{-1}. Therefore, decarboxylation constitutes the rate-limiting step in pathway II. By comparing pathways I and II, we

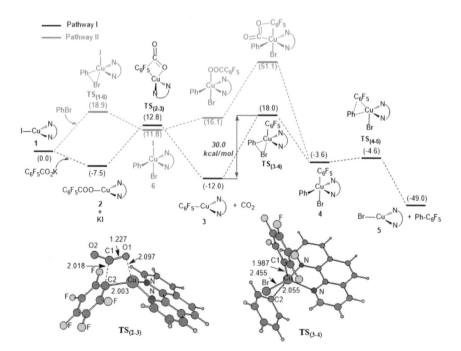

Fig. 3.1 Comparing two plausible mechanisms for Cu-catalyzed decarboxylative cross-coupling (B3LYP method. SDD basis set for I, 6-31G(d) for C, H, O, F, Cu, and Br. Solvation = CPCM/bondi). Reproduced from Angew. Chem. Int. Ed. 2009, 48, 9350 by permission of John Wiley & Sons Ltd

conclude that decarboxylation likely occurs on copper(I) before oxidative addition. This conclusion is consistent with the observation by Sheppard et al. [23a, b].

3.3 Conclusion

In summary, the decarboxylative cross-coupling of potassium polyfluorobenzoates with aryl iodides and bromides mediated by a copper-only system was discovered. This reaction represents both a new type of copper-catalyzed cross-coupling reaction and a new type of transition metal-catalyzed decarboxylative coupling reaction. The reaction is practical for the preparation of polyfluorobiaryls from readily accessible and nonvolatile polyfluorobenzoate salts. In contrast to the previously reported decarboxylative coupling reactions, palladium is not required for the present transformation, meaning that both decarboxylation and cross-coupling steps are catalyzed solely by copper in the newly discovered process. Theoretical analyses suggest that decarboxylation should occur at first on copper(I) to generate a polyfluorophenylcopper(I) intermediate, which then reacts with aryl halides through oxidative addition and reductive elimination to produce the coupling products.

3.4 Experimental Section and Compound Data

3.4.1 General Information

All reactions were carried out in oven-dried Schlenk tube under Argon atmosphere (purity $\geq$ 99.999%). The diglyme solvent was bought from Alfa Aesar without further purification. Dimethylacetamide (DMA) solvent was bought from Sinopharm Chemical Reagent Co. Ltd and dried on 4 Å molecular sieve, degassed with Argon before use. All aryl halides were bought from Alfa Aesar or Acros and used as is. All the perfluorobenzoic acids were purchased from Sigma Aldrich. Copper(I) iodide was purchased from Sinopharm Chemical Reagent Co. Ltd as an off-white powder and refluxed in THF overnight using a Soxhlet-Dean-Stark extractor for further purification. Anhydrous 1,10-phenanthroline was bought from Acros. All the other reagents and solvents mentioned in this text were bought from Sinopharm Chemical Reagent Co. Ltd or Alfa Aesar and purified when necessary.

^{1}H-NMR, ^{13}C-NMR, ^{19}F-NMR spectra were recorded on a Bruker Avance 400 spectrometer at ambient temperature in CDCl$_3$ unless otherwise noted. Data for ^{1}H-NMR are reported as follows: chemical shift (δ m), multiplicity, integration, and coupling constant (Hz). Data for ^{13}C-NMR are reported in terms of chemical shift (δ ppm), multiplicity, and coupling constant (Hz). ^{19}F-NMR were recorded in CDCl$_3$ solutions and trifluorotoluene (TFT) ($\delta = -67.73$ ppm) was employed as an external standard. Data for ^{19}F-NMR are reported as follows: chemical shift (δ m), multiplicity, integration, and coupling constant (Hz). Gas chromatographic (GC) analysis was acquired on a Shimadzu GC-2014 Series GC System equipped with a flame ionization detector. GC-MS analysis was performed on Thermo Scientific AS 3000 Series GC-MS System. HRMS analysis was performed on Finnigan LCQ advantage Max Series MS System. Elementary Analysis was carried out on Elementar Vario EL III elemental analyzer. Organic solutions were concentrated under reduced pressure on a Buchi rotary evaporator. Flash column chromatographic purification of products was accomplished using forced-flow chromatography on Silica Gel (200–300 mesh).

3.4.2 General Procedure

The coupling of potassium pentafluorobenzoate with various aryl iodides:

Copper(I) iodide (0.05 mmol, 9.5 mg) appointed amount of potassium pentafluorobenzoate (0.6–0.75 mmol, See Part 3) and the aryl iodides (0.50 mmol) (if solid) were placed in an oven-dried 10 ml Schlenk tube, the reaction vessel was evacuated and filled with argon for three times. Then aryl iodide (0.50 mmol) (if liquid) and diglyme (0.5 ml) were added with a syringe under a counter flow of argon, the vessel was sealed with a screw cap, stirred at room temperature for 10 min, and connected to the Schlenk line which was full of argon, stirred at 130 °C for the appointed time. Upon completion of the reaction, the mixture was cooled to room temperature and diluted with ethyl acetate or petroleum ether (20 ml) and filtered through a short silica gel column to remove the deposition. The organic layers were washed with water (20 ml × 3), and then with brine, dried over Na_2SO_4, and filtered, the solvents were removed via rotary vapor. Purification of the residue by column chromatography (silica gel, ethyl acetate/hexane gradient) yielded the corresponding fluoroarene.

3.4.3 Characterization of the Products

Compound name 2,3,4,5,6-pentafluorobiphenyl

122 mg (99%) of a white solid was obtained. This compound is known. Reference: Korenaga, T.; Kosaki, T.; Fukumura, R.; Ema, T.; Sakai, T. Org. Lett. 2005, 7, 4915–4918. ^{1}H-NMR (400 MHz, CDCl$_3$, 293 K, TMS): δ 7.39–7.42 (m, 2H), 7.44–7.48 (m, 3H). ^{13}C-NMR (100 MHz, CDCl$_3$, 293 K, TMS): δ 115.9 (td, J_F = 3.8, 17.5 Hz), 126.4, 128.7, 129.3, 130.1, 137.8 (dm, J_F = 253.2 Hz), 140.4

(dm, J_F = 253.2 Hz), 144.2 (dm, J_F = 252.4 Hz). ^{19}F-NMR (377 MHz, CDCl$_3$, 293 K, TFT): −162.3 (m, 2F), −155.6 (t, J_F = 21.1 Hz, 1F), −143.2 (dd, J_F = 23.0, 8.3 Hz, 2F).

Compound name ethyl 2′,3′,4′,5′,6′-pentafluorobiphenyl-4-carboxylate

158 mg (99%) of a white solid was obtained. This compound is known. Reference: Lafrance, M.; Rowley, C. N.; Woo, T. K.; Fagnou, K. J. Am. Chem. Soc. 2006, 128, 8754–8756. ^{1}H-NMR (400 MHz, CDCl$_3$, 293 K, TMS): δ 1.42 (t, J = 7.2 Hz, 3H), 4.42 (q, J = 7.2 Hz, 2H), 7.50–7.52 (m, 2H), 8.16–8.18 (m, 2H). ^{13}C-NMR (100 MHz, CDCl$_3$, 293 K, TMS): δ 14.2, 61.2, 115.0 (td, J_F = 4.2, 17.6 Hz), 129.8, 130.2, 130.8, 131.3, 137.9 (dm, J_F = 251.4 Hz), 140.8 (dm, J_F = 253.3 Hz), 144.1 (dm, J_F = 247.4 Hz), 165.8. ^{19}F-NMR (377 MHz, CDCl$_3$, 293 K, TFT): −161.6 (m, 2F), −154.2 (t, J_F = 20.7 Hz, 1F), −142.7 (dd, J_F = 22.2, 8.3 Hz, 2F).

Compound name 2,3,4,5,6-pentafluoro-4′-(trifluoromethyl)biphenyl

156 mg (99%) of a white solid was obtained. This compound is known. Reference: Lafrance, M.; Rowley, C. N.; Woo, T. K.; Fagnou, K. J. Am. Chem. Soc. 2006, 128, 8754–8756. ^{1}H-NMR (400 MHz, CDCl$_3$, 293 K, TMS): δ 7.55−7.57 (m, 2H), 7.76 −7.78 (m, 2H). ^{13}C-NMR (100 MHz, CDCl$_3$, 293 K, TMS): δ 114.6 (td, J_F = 3.9, 16.5 Hz), 123.8 (q, J_F = 270.7 Hz), 125.7(q, J_F = 3.6 Hz), 130.2, 130.7, 131.6 (q, J_F = 32.6 Hz), 138.0 (dm, J_F = 251.6 Hz), 141.0 (dm, J_F = 253.8 Hz), 144.2 (dm, J_F = 247.5 Hz). ^{19}F-NMR (377 MHz, CDCl$_3$, 293 K, TFT): −161.5 (m, 2F), −153.8 (t, J_F = 20.7 Hz, 1F), −142.9 (dd, J_F = 21.9, 8.3 Hz, 2F), −63.0 (s, 3F).

Compound name 2-(perfluorophenyl)pyridine

122 mg (99%) of a pale yellow solid was obtained. This compound is known. Reference: Do, H. Q.; Daugulis, O. J. Am. Chem. Soc. 2008, 130, 1128–1129. ^{1}H-NMR (400 MHz, CDCl$_3$, 293 K, TMS): δ 7.37–7.40 (m, 1H), 7.46–7.49 (m, 1H), 7.82–7.86 (m, 1H), 8.76–8.78 (m, 1H). ^{13}C-NMR (100 MHz, CDCl$_3$, 293 K, TMS): δ 115.4 (td, J_F = 3.8, 16.8 Hz), 123.8, 125.9, 136.6, 137.7 (dm, J_F = 251.1 Hz), 141.2 (dm, J_F = 253.5 Hz), 144.6 (dm, J_F = 244.7 Hz), 146.8, 150.1. ^{19}F-NMR (377 MHz, CDCl$_3$, 293 K, TFT): −161.9 (m, 2F), −153.8 (t, J_F = 20.4 Hz, 1F), −143.2 (dd, J_F = 23.4, 7.5 Hz, 2F).

Compound name 2,2′,3,4,4′,5,6-heptafluorobiphenyl

132 mg (94%) of a white solid was obtained. This compound is new. ^{1}H-NMR (400 MHz, CDCl$_3$, 293 K, TMS): δ 6.97–7.05 (m, 2H), 7.31–7.36 (m, 1H). ^{13}C-NMR (100 MHz, CDCl$_3$, 293 K, TMS): δ 104.7 (t, J_F = 25.4 Hz), 115.5 (td, J_F = 3.4, 18.4 Hz), 110.4 (dd, J_F = 1.8, 16.7 Hz), 112.0 (dd, J_F = 3.6, 21.7 Hz), 132.9 (d, J_F = 9.6 Hz), 137.8 (dm, J_F = 251.3 Hz), 141.3 (dm, J_F = 253.5 Hz), 144.4 (dm, J_F = 247.7 Hz), 160.4 (d, J_F = 251.7 Hz), 164.1 (d, J_F = 251.0 Hz). ^{19}F-NMR (377 MHz, CDCl$_3$, 293 K, TFT): −161.7 (m, 2F), −153.6 (t, J_F = 20.4 Hz, 1F), −140.3 (m, 2F), −108.1 (dd, J_F = 1.0, 20.4 Hz, 1F), −106.7 (d, J_F = 8.7 Hz, 1F). HRMS calcd for C$_{12}$H$_3$F$_7$ (M+) 280.0123; found: 280.0121.

Compound name 6-(perfluorophenyl)-2,3-dihydrobenzo[b][1,4]dioxine

151 mg (99%) of a tan solid was obtained. This compound is new. ^{1}H-NMR (400 MHz, CDCl$_3$, 293 K, TMS): δ 4.31 (t, J = 0.8, Hz 4H), 6.88–6.91 (m, 1H), 6.94–6.98 (m, 2H). ^{13}C-NMR (100 MHz, CDCl$_3$, 293 K, TMS): δ 64.3, 64.4, 115.5 (td, J_F = 3.8, 16.8 Hz), 117.5, 119.1, 123.3, 123.5, 137.8 (dm, J_F = 250.4 Hz), 140.1 (dm, J_F = 252.0 Hz), 144.2 (dm, J_F = 245.4 Hz), 143.6, 144.5. ^{19}F-NMR (377 MHz, CDCl$_3$, 293 K, TFT): −162.5 (m, 2F), −156.3 (t, J_F = 21.1 Hz, 1F), −143.3 (dd, J_F = 23.3, 8.3 Hz, 2F). HRMS calcd for C$_{14}$H$_7$F$_5$O$_2$ (M+) 302.0366; found: 302.0361.

Compound name 2-nitro-5-(perfluorophenyl)thiophene

90 mg (61%) of a yellow solid was obtained. This compound is new. ^{1}H-NMR (400 MHz, CDCl$_3$, 293 K, TMS): δ 7.47–7.48 (m, 1H), 7.96–7.97 (m, 1H). ^{13}C-NMR (100 MHz, CDCl$_3$, 293 K, TMS): δ 107.9 (td, J_F = 4.4, 15.0 Hz), 128.1, 129.4, 133.4, 138.0 (dm, J_F = 249.5 Hz), 141.4 (dm, J_F = 257.3 Hz), 144.5 (dm, J_F = 244.3 Hz), 153.3. ^{19}F-NMR (377 MHz, CDCl$_3$, 293 K, TFT): −160.1 (m, 2F), −150.9 (t, J_F = 20.7 Hz, 1F), −138.3 (dd, J_F = 20.0, 4.8 Hz, 2F). HRMS calcd for C$_{10}$H$_2$F$_5$NO$_2$S (M+) 294.9726; found: 294.9730.

Compound name 2,6-difluoro-4′-(trifluoromethyl)biphenyl

56 mg (86%) of a white solid was obtained. This compound is new. ^{1}H-NMR (400 MHz, CDCl$_3$, 293 K, TMS): δ 6.98–7.05 (m, 2H), 7.30–7.307 (m, 1H), 7.59 (d, J = 8.4, 2H), 7.72 (d, J = 8.4, 2H). ^{13}C-NMR (100 MHz, CDCl$_3$, 293 K, TMS): δ 111.8 (dd, J_F = 6.6, 19.3 Hz), 117.2 (t, J_F = 18.3 Hz), 124.1(q, J_F = 270.5 Hz), 125.2 (q, J_F = 3.8 Hz), 129.8 (t, J_F = 10.2 Hz), 130.3 (q, J_F = 32.3 Hz), 130.7, 133.0, 160.0 (dd, J_F = 6.8, 248.1 Hz). ^{19}F-NMR (377 MHz, CDCl$_3$, 293 K, TFT): −114.5 (s, 2F), −62.8 (s, 3F). HRMS calcd for C$_{13}$H$_7$F$_5$ (M+) 258.0468; found: 258.0464.

Compound name 3-(2,6-difluorophenyl)pyridine

42 mg (88%) of a yellow oil was obtained. This compound is new. ^{1}H-NMR (400 MHz, CDCl$_3$, 293 K, TMS): δ 7.02 (t, J = 8.4 Hz, 2H), 7.31–7.38 (m, 1H), 7.41 (dd, J = 4.8, 7.6 Hz, 2H), 7.81 (d, J = 8.0 Hz, 1H), 8.64 (d, J = 4.0 Hz, 1H), 8.73 (s, 1H). ^{13}C-NMR (100 MHz, CDCl$_3$, 293 K, TMS): δ 111.8 (dd, J_F = 6.7, 19.3 Hz), 115.0 (t, J_F = 18.4 Hz), 123.3, 125.6, 129.9 (t, J_F = 10.3 Hz), 137.8, 149.0, 150.6, 160.1 (dd, J_F = 6.7, 248.2 Hz). ^{19}F-NMR (377 MHz, CDCl$_3$, 293 K, TFT): −114.5 (s, 2F). HRMS calcd for C$_{11}$H$_7$F$_2$N (M+) 191.0547; found: 191.0540.

Compound name 2,3,5,6-tetrafluoro-4′-methoxy-4-methylbiphenyl

66 mg (97%) of a white crystal was obtained. This compound is new. ^{1}H-NMR (400 MHz, CDCl$_3$, 293 K, TMS): δ 2.31 (t, J_F = 2.0 Hz, 3H), 3.86 (s, 3H), 7.00 (d, J = 8.8 Hz, 2H), 7.39 (d, J = 8.8 Hz, 2H). ^{13}C-NMR (100 MHz, CDCl$_3$, 293 K, TMS): δ 7.44, 55.2, 114.0, 114.4 (t, J_F = 19.1 Hz), 117.6 (t, J_F = 16.5 Hz), 119.8, 131.4, 143.7 (dm, J_F = 243.3 Hz), 145.3 (dm, J_F = 242.3 Hz), 160.0. ^{19}F-NMR (377 MHz, CDCl$_3$, 293 K, TFT): −146.0 (dd, J_F = 12.4, 22.2 Hz, 2F), −144.4 (dd, J_F = 12.8, 22.6 Hz, 2F). HRMS calcd for C$_{14}$H$_{10}$F$_4$O (M+) 270.0668; found: 270.0666.

Compound name ethyl 2′,3′,5′,6′-tetrafluoro-4′-methylbiphenyl-4-carboxylate

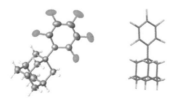

74 mg (95%) of a yellow solid was obtained. This compound is new. ^{1}H-NMR (400 MHz, CDCl$_3$, 293 K, TMS): δ 1.42 (t, J = 7.2 Hz, 3H), 2.33 (t, J_F = 2.0 Hz, 3H), 4.41 (q, J = 7.2 Hz, 2H), 7.53 (d, J = 8.0 Hz, 2H), 8.15 (d, J = 8.4 Hz, 2H). ^{13}C-NMR (100 MHz, CDCl$_3$, 293 K, TMS): δ 7.6, 14.3, 61.2, 115.9 (t, J_F = 20.0 Hz), 117.0 (t, J_F = 16.4 Hz), 129.6, 130.2, 130.8, 132.2, 143.5 (dm, J_F = 245.4 Hz), 145.3 (dm, J_F = 243.1 Hz), 166.0. ^{19}F-NMR (377 MHz, CDCl$_3$, 293 K, TFT): −145.2 (dd, J_F = 12.4, 21.5 Hz, 2F), −143.6 (dd, J_F = 13.2, 22.6 Hz, 2F). HRMS calcd for C$_{16}$H$_{12}$F$_4$O$_2$ (M+) 312.0773; found: 312.0768.

Compound name 1-(perfluorophenyl)adamantane

148 mg (98%) of a white crystalline solid was obtained. This compound is known.
    Reference: Cairncross; Sheppard. J. Am. Chem. Soc. 1968, 90, 2186. ^{1}H-NMR (400 MHz, CDCl$_3$, 293 K, TMS): δ 1.75–1.83 (m, 6H), 2.01 (s, 3H), 2.22 (s, 6H). ^{13}C-NMR (100 MHz, CDCl$_3$, 293 K, TMS): δ 28.9, 36.5, 40.5, 41.1 (t, J_F = 5.2 Hz), 122.3–122.6 (m), 137.9 (dm, J_F = 246.0 Hz), 139.0 (dm, J_F = 250.2 Hz), 146.3 (dm, J_F = 247.3 Hz). ^{19}F-NMR (377 MHz, CDCl$_3$, 293 K, TFT): −163.0 (m, 2F), −158.5 (t, J_F = 21.9 Hz, 1F), −138.1 (m, 2F).

Crystal data and structure

Single crystals of C$_{32}$H$_{30}$F$_{10}$ were recrystallised from **CHCl$_3$** mounted in inert oil and transferred to the cold gas stream of the diffractometer.

Crystal structure determination of 1-(perfluorophenyl)adamantane
Crystal Data. C$_{32}$H$_{30}$F$_{10}$, M = 302.16, Triclinic, a = 6.535(5) Å, b = 10.087(5) Å, c = 11.094(5) Å, α = 106.463(5)°, β = 105.428(5)°, γ = 98.691(5)°, U = 655.4(7) Å^3, T = 295(2), space group P-1 (no. 2), Z = 2, μ(Mo-Kα) = 0.138, 7264 reflections measured, 2844 unique (R_{int} = 0.0166) which were used in all calculations. The final $wR(F_2)$ was 0.0590 (all data).

References

1. (a) Baudoin, O. (2007). *Angewandte Chemie International Edition*, *46*, 1373–1375. (b) Goossen, L. J., Rodriguez, N., & Goossen, K. (2008). *Angewandte Chemie International Edition*, *47*, 3100–3120.
2. (a) Nilsson, M. (1966). *Acta Chemica Scandinavica*, *20*, 423–907. (b) Björklund, C., & Nilsson, M. (1968). *Acta Chemica Scandinavica*, *22*, 2585–2588. (c) Cairncross, A., Roland, J. R., Henderson, R. M., & Shepard, W. A. (1970). *Journal of the American Chemical Society*, *92*, 3187–3189.
3. (a) Goossen, L. J., Deng, G., & Levy, L. M. (2006). *Science*, *313*, 662–664. (b) Goossen, L., Rodriguez, J.N., Melzer, B., Linder, C., Deng, G., & Levy, L. M. (2007). *Journal of the American Chemical Society*, *129*, 4824–4833. (c) Goossen, L. J., & Melzer, B. (2007). *The Journal of Organic Chemistry*, *72*, 7473–7476. (d) Goossen, L. J., & Knauber, T. (2008). *The Journal of Organic Chemistry*, *73*, 8631–8634. (e). Goossen, L. J., Zimmermann, B., & T. Knauber. (2008). *Angewandte Chemie International Edition*, *47*, 7103–7106. (f) Goossen, L. J., Rodriguez, N., & Linder, C. (2008). *Journal of the American Chemical Society*, *130*, 15248–15249. (g) Goossen, L. J., Rudolphi, F., Oppel, C., & Rodriguez, N. (2008). *Angewandte Chemie International Edition*, *47*, 3043–3045.
4. (a) Myers, A. G., Tanaka, D., & Mannion, M. R. (2002). *Journal of the American Chemical Society*, *124*, 11250–11251. (b) Tanaka, D., & Myers, A. G. (2004). *Organic Letters*, *6*, 433–436. (c) Tanaka, D., Romeril, S. P., & Myers, A. G. (2005). *Journal of the American Chemical Society*, *127*, 10323–10333.
5. Forgione, P., Brochu, M.-C., St-Onge, M., Thesen, K. H., Bailey, M. D., & Bilodeau, F. (2006). *Journal of the American Chemical Society*, *128*, 11350–11351.
6. (a) Moon, J., Jeong, M., Nam, H., Ju, J., Moon, J. H., Jung, H. M., & Lee, S. (2008). *Organic Letters*, *10*, 945–948. (b) Moon, J., Jang, M., & Lee, S. J. (2009). *Organic Letters*, *74*, 1403–1408.
7. (a) Becht, J.-M., & Le Drian, C. (2008). *Organic Letters*, *10*, 3161–3164. (b) Becht, J.-M., Catala, C., Le Drian, C., & Wagner, A. (2007). *Organic Letters*, *9*, 1781–1784.
8. (a) Wang, Z. Y., Ding, Q. P., He, X. D., & Wu, J. (2009). *Organic & Biomolecular Chemistry*, *7*, 863–864. (b) Wang, Z. Y., Ding, Q. P., He, X. D., & Wu, J. (2009). *Tetrahedron*, *65*, 4635–4638. (c) Voutchkova, A., Coplin, A., Leadbeater, N. E., & Crabtree, R. H. (2008). *Chemical Communications*, 6312–6313.
9. Wang, C., Piel, I., & Glorius, F. (2009). *The Journal of Organic Chemistry*, *131*, 4194–4200.
10. Shang, R., Fu, Y., Li, J. B., Zhang, S. L., Guo, Q. X., & Liu, L. (2009). *The Journal of Organic Chemistry*, *131*, 5738–5743.
11. Waetzig, S. R., Rayabarupu, D. K., Weaver, J. D., & Tunge, J. A. (2006). *Angewandte Chemie International Edition*, *45*, 4977–4985.
12. There are two literature examples for palladium-catalyzed decarboxylative coupling of perfluorobenzoic acid with 4-iodoanisole; see references [7b, 8c].
13. For a special protocol for the Suzuki–Miyaura coupling reaction of pentafluorophenylboronic acid, which is an inactive substrate under normal conditions, see: Korenaga, T., Kosaki, T., Fukumura, R., Ema, T., & Sakai, T. (2005). *Organic Letters*, *7*, 4915–4925.
14. (a) Zahn, A., Brotschi, C., & Leumann, C. (2005). *Chemistry—A European Journal*, *11*, 2125–2129. (b) Mewshaw, R. E., Edsall Jr., R. J., Yang, C., Manas, E. S., Xu, Z. B., Henderson, R. A., Keith Jr., J. C., & Harris, H. A. (2005). *Journal of Medicinal Chemistry*, *48*, 3953–3959. (c) de Candia, M., Liantonio, F., Carotti, A., De Cristofaro, R., & Altomare, C. (2009). *Journal of Medicinal Chemistry*, *52*, 1018–1028.
15. (a) Sakamoto, Y., Suzuki, T., Miura, A., Fujikawa, H., Tokito, S., & Taga, Y. (2000). *Journal of the American Chemical Society*, *122*, 1832–1833. (b) Nitschke, J. R., & Tilley, T. D. (2001). *Journal of the American Chemical Society*, *123*, 10183–10190. (c) Zacharias, P., Gather, M. C., Rojahn, M., Nuyken, O., & Meerholz, K. (2007). *Angewandte Chemie International Edition*, *46*, 4388–4392.

16. (a) Lafrance, M., Rowley, C. N., Woo, T. K., & Fagnou, K. (2006). *Journal of the American Chemical Society, 128*, 8754–8756. (b) Lafrance, M.; Shore, D., & Fagnou, K. (2006). *Organic Letters, 8*, 5097–5100.
17. (a) Do, H.-Q., & Daugulis, O. (2007). *Journal of the American Chemical Society, 129*, 12404–12405. (b) Do, H.-Q., & Daugulis, O. (2008). *Journal of the American Chemical Society, 130*, 1128–1129. (c) Do, H.-Q., Kashif Khan, R. M., & Daugulis, O. (2008). *Journal of the American Chemical Society, 130*, 15185–15192.
18. Goossen, L. J., Manjolinho, F., Khan, B. A., & Rodriguez, N. J. (2009). *Organic Letters, 74*, 2620–2623.
19. Reviews for copper-catalyzed cross-couplings : (a) Ley, S. V., & Thomas, A. W. (2003). *Angewandte Chemie International Edition, 42*, 5400–5449. (b) Deng, W., Liu, L., & Guo, Q.-X. (2004). *Chinese Journal of Organic Chemistry, 24*, 150–165. (c) Monnier, F., & Taillefer, M. (2008). *Angewandte Chemie International Edition, 47*, 3096–3099. (d) Evano, G., Blanchard, N., & Toumi, M. (2008). *Chemical Reviews, 108*, 3054–3131.
20. For selected examples of copper-catalyzed cross-couplings, see: (a) Klapars, A., Antilla, J. C., Huang, X., & Buchwald, S. L. (2001). *Journal of the American Chemical Society, 123*, 7727–7729. (b) Shafir, A., & Buchwald, S. L. (2006). *Journal of the American Chemical Society, 128*, 8742–8743. (c) Altman, R. A., Hyde, A. M., Huang, X., & Buchwald, S. L. (2008). *Journal of the American Chemical Society, 130*, 9613–9620. (d) Ma, D., Zhang, Y., Yao, J., Wu, S., & Tao, F. (1998). *Journal of the American Chemical Society, 120*, 12459–12467. (e) Ma, D., & Liu, F. (2004). *Chemical Communications*, 1934–1939. (f) Ma, D., Xie, S., Xue, P., Zhang, X., Dong, J., & Jiang, Y. (2009). *Angewandte Chemie International Edition, 48*, 4222–4225. (g) Thathagar, M. B., Beckers, J., & Rothenberg, G. (2002). *Journal of the American Chemical Society, 124*, 11858–11859. (h) del Amo, V., Dubbaka, S. R., Krasovskiy, A., & Knochel, P. (2006). *Angewandte Chemie International Edition, 45*, 7838–7842. (i) Chen, X., Hao, X.-S., Goodhue, C. E., & Yu, J.-Q. (2006). *Journal of the American Chemical Society, 128*, 6790–6791. (j) Villalobos, J. M., Srogl, J., & Libeskind, L. S. (2007). *Journal of the American Chemical Society, 129*, 15734–15735. (k) Prokopcova, H., & Kappe, C. O. (2008). *Angewandte Chemie International Edition, 47*, 3674–3676. (l) Fuller, P. H., Kim, J.-W., & Chemler, S. R. (2008). *Journal of the American Chemical Society, 130*, 17638–17639. (m) King, A. E., Brunold, T. C., & Stahl, S. S. (2009). *Journal of the American Chemical Society, 131*, 5044–5045. (n) Phipps, R. J., & Gaunt, M. J. (2009). *Science, 323*, 1593–1601.
21. Goossen, L. J., Thiel, W. R., Rodriguez, N., Linder, C., & Melzer, B. (2007). *Advanced Synthesis & Catalysis, 349*, 2241–2252.
22. Some previous studies on the mechanism of copper-catalyzed cross-couplings: (a) Ouali, A., Spindler, J.-F., Jutand, A., & Taillefer, M. (2007). *Advanced Synthesis & Catalysis, 349*, 1906–1913. (b) Zhang, S.-L., Liu, L., Fu, Y., & Guo, Q.-X. (2007). *Organometallics, 26*, 4546–4551. (c) Tye, J. W., Weng, Z., Johns, A. M., Incarvito, C. D., & Hartwig, J. F. (2008). *Journal of the American Chemical Society, 130*, 9971–9983. (d) Huffman, L. M., & Stahl, S. S. (2008). *Journal of the American Chemical Society, 130*, 9196–9197. (e) Kaddouri, H., Vicente, V., Ouali, A.; Ouazzani, F., & Taillefer, M. (2009). *Angewandte Chemie International Edition, 48*, 333–346. (f) Strieter, E. R., Bhayana, B., & Buchwald, S. L. (2009). *Journal of the American Chemical Society, 131*, 78–88.
23. Earlier studies already showed that pentafluorophenylcopper can react with aryl iodides producing coupling products in high yield. See: (a) Cairncross, A., & Sheppard, W. A. (1968). *Journal of the American Chemical Society, 90*, 2186–21187. (b) Sheppard, W. A. (1970). *Journal of the American Chemical Society, 92*, 5419–5422.

Chapter 4
Palladium-Catalyzed Decarboxylative Couplings of Potassium Polyfluorobenzoates with Aryl Bromides, Chlorides, and Triflates

Abstract Pd-catalyzed decarboxylative cross-coupling of potassium polyfluorobenzoates with aryl bromides, chlorides, and triflates is achieved using diglyme as the solvent. The reaction is useful for synthesis of polyfluorobiaryls from readily accessible and nonvolatile polyfluorobenzoate salts. Different from the Cu-catalyzed decarboxylation cross-coupling where oxidative addition is the rate-limiting step, in the Pd-catalyzed version decarboxylation is the rate-limiting step.

4.1 Introduction

Consideration the guidelines of green chemistry as well as the synthetic efficiency, transition metal-catalyzed decarboxylative coupling reactions that use carboxylic acids as aryl sources have unique advantages [1]. Decarboxylaitve couplings avoid the use of expensive and/or sensitive organometallic reagents and generate CO_2 without producing toxic metal halides. Goossen et al. showed the decarboxylative coupling of some benzoic acids and α-oxo carboxylates with aryl halides and triflates could be achieved through a Pd/Cu bimetallic catalysis [2]. Myers [3], Forgione [4], and other groups [5] found that Pd could catalyze the decarboxylative cross-coupling of some carboxylic acids with olefins and aryl iodides or bromides. Other recent studies by Tunge, Li, Miura, Chruma, and other groups also highlighted the synthetic utility of related decarboxylative reactions [6].

© Springer Nature Singapore Pte Ltd. 2017
R. Shang, *New Carbon–Carbon Coupling Reactions Based on Decarboxylation and Iron-Catalyzed C–H Activation*, Springer Theses,
DOI 10.1007/978-981-10-3193-9_4

In the previous chapter, we described that a Cu-only catalyst system is capable of catalyzing the decarboxylative coupling of potassium polyfluorobenzoates with aryl iodides and bromides [7]. Here we report that a catalyst system based on solely palladium, which can also catalyze the similar decarboxylative cross-couplings, but importantly, with aryl bromides, chlorides, and even triflates. Resembling the Cu-catalyzed versions [7], the newly developed Pd-catalyzed reactions can also replace the use of expensive but often less reactive [8] fluorobenzene organometallics in the synthesis of polyfluorobiaryls, which are of interest in materials science [9] and medicinal chemistry [10]. The new reactions also provide a practical method complementary to direct arylation reported independently by Fagnou [11] and Daugulis [12] fluorobiaryl synthesis through direct C–H arylation of polyfluoroarene. Furthermore, through theoretical analysis we revealed that the palladium-catalyzed decarboxylative cross-coupling is distinct in the mechanism compared with the similar reaction catalyzed by copper catalyst.

4.1.1 Investigation of the Reaction Conditions

Our work started with the decarboxylative coupling of potassium pentafluorobenzoate with o-tolyl chlorides. We chose o-tolyl chloride as a model substrate aiming to optimize the reaction with tolerance to steric effect. Aryl chlorides are cheap and relatively inert compared with corresponding iodide and bromide and it cannot be used in the copper-catalyzed reaction discussed in the previous chapter. It should be noted there were two examples in literatures [5d, 5g] for Pd-catalyzed decarboxylative coupling of pentafluorobenzoic acid with 4-iodoanisole. However, the aryl bromides, chlorides, and triflates have not successfully used in this reaction. After extensive experimentation, (Table 4.1) it was found that the coupling reaction did not proceed well in many solvents (entries 1–4), while it worked smoothly when diglyme was used as solvent. The reaction took place readily (yield 88%) using a simple ligand (PCy₃) in diglyme solvent (entry 5). The use of more popular Ar–Cl activation ligands such as ᵗBu₃P, S-Phos, and Dave-Phos [13] did not produce a better result (entries 6–9), and the use of other palladium salts (entries 10–12) gave similar results to the reaction using Pd(OAc)₂. Decreasing the loading of palladium catalyst to 1 mol% (entry 13) resulted in a slightly lower yield of 86%. Moreover, the same catalyst system can also be applied to o-tolylbromide (entry 14). It is interesting that a less electron-rich ligand (P(o-Tol)₃) is more effective for o-tolylbromide than PCy₃ (entry 15).

4.1.2 Exploration of the Substrate Scope

Extending the model reaction to various substrates showed that both electron-rich and electron-poor aryl bromides and chlorides can be successfully converted,

Table 4.1 Pd-catalyzed decarboxylative cross-coupling of potassium Pentafluorobenzoate with o-Tolylhalide[a]

Entry	X	Pd source	Ligand	Solvent	Yield%[b]
1	Cl	Pd(OAc)$_2$	P(Cy)$_3$	NMP	Trace
2	Cl	Pd(OAc)$_2$	P(Cy)$_3$	DMA	Trace
3	Cl	Pd(OAc)$_2$	P(Cy)$_3$	DMF	11
4	Cl	Pd(OAc)$_2$	P(Cy)$_3$	Mesitylene	n.r.
5	Cl	Pd(OAc)$_2$	P(Cy)$_3$	Diglyme	88
6	Cl	Pd(OAc)$_2$	P(t-Bu)$_3$	Diglyme	79
7	Cl	Pd(OAc)$_2$	S-Phos	Diglyme	85
8	Cl	Pd(OAc)$_2$	Dave-Phos	Diglyme	28
9	Cl	Pd(OAc)$_2$	JohnPhos	Diglyme	15
10	Cl	Pd$_2$(dba)$_3$	P(Cy)$_3$	Diglyme	84
11	Cl	Pd(TFA)$_2$	P(Cy)$_3$	Diglyme	85
12	Cl	Pd(MeCN)$_2$Cl$_2$	P(Cy)$_3$	Diglyme	73
13[c]	Cl	Pd(OAc)$_2$	P(Cy)$_3$	Diglyme	86
14	Br	Pd(OAc)$_2$	P(Cy)$_3$	Diglyme	87
15	Br	Pd(OAc)$_2$	P(o-Tol)$_3$	Diglyme	91

[a]All the reactions were carried out at 0.25 mmol scale in 0.5 ml solvents. [b]GC yields using biphenyl as the internal standard. [c]1 mol% Pd(OAc)$_2$ and 2 mol% ligand were used

tolerating a range of functional groups (Table 4.2). In several cases (entries 1, 4, and 15), aryl triflates (but not aryl tosylates, see entry 13) can also be successfully converted. Ortho-substitution can be tolerated in the transformation (entries 3, 5, 7, and 8). It is worth mentioning that the desired product can be obtained in 81% for 2-bromo-m-xylene which has large ortho steric hindrance. In addition, some heteroaryl bromides and chlorides can be used to produce the corresponding polyfluorobiaryls (entries 21 and 22). The yield of 3-bromopyridine was 81%, but for 2-bromopyridine the reaction did not take place possible due to catalyst poisoning by the substrate. The yield of 2-Chlorothiophene is 75%, while the yield of 3-chlorothiophene is only moderate (52%).

The scope of the reaction with respect to fluoroarene is shown in Table 4.3. Considering the different propensities for decarboxylation of benzoate with different fluorine substitutions, a higher loading of palladium catalyst (4 mol%) was applied in the scope studies in Table 4.3. Under the optimized conditions, potassium

Table 4.2 Pd-catalyzed decarboxylative cross-coupling of C_6F_5COOK with aryl bromides, chlorides, and triflates[a]

Entry	Sbustrate		Yield[b]	Entry	Sbustrate		Yield[b]
1		Cl	91	12		Cl	86
		Br	89				
		OTf	76				
2		Cl	96	13		Cl	96
		Br	92			Br	92
3		Cl	84	14		Br	93
		Br	86				
4		OTf	79	15		OTf	89
		Cl	85			Cl	97
5		Cl	94	16		Cl	83
6		Cl	96	17		Cl	99
7		Br	81	18		Cl	97
8		Cl	83	19		Cl	98
9		Br	93	20		Cl	95
10		Cl	89	21		3-Br	81
		Br	84			2-Br	trace
11		Br	87	22		2-Cl	75
						3-Cl	52

[a]All the reactions were carried out at 0.25 mmol scale in 0.5 ml of diglyme. [b]Isolated yields were calculated based on the quantity of aryl halide

Table 4.3 Pd-catalyzed decarboxylative cross-coupling of various potassium polyfluoroben-zoates[a]

Entry	Product		Yield[b]	Entry	Product		Yield[b]
1	(structure)	Cl	trace	7	(structure)	Br	73
2	(structure)	Br	68	8	(structure)	Cl	trace
3	(structure)	Cl	81	9[c]	(structure)	Cl	14
		Br	74				
4	(structure)	Cl	85	10[c]	(structure)	Cl	31
		Br	82				
5	(structure)	Cl	96	11	(structure)	Cl	89
		Br	92			Br	96
6	(structure)	Cl	51	12	(structure)	Cl	85

[a]All the reactions were carried out at 0.25 mmol scale in 0.5 ml diglyme. [b]Isolated yields were calculated based on the quantity of aryl halide. [c]Di- and tri-arylations were observed

monofluorobenzoate cannot be efficiently converted, unless an *ortho*-Cl or *ortho*-CF$_3$ group is presented (entries 1–3). 2-Chloro-6-fluorobenzoate can react to deliver desired product in 68% yield (entry 2). In the presence of two *ortho*-fluorine substituents, decarboxylative coupling of potassium perfluorobenzoate proceeded smoothly with both 4-methoxyphenyl bromide and chloride (entries 4–7). The reactions were also successful when tri- and tetrafluorobenzoates with two *ortho*-F atoms were applied (entries 9–12), while unsuccessful for substrate possessing only one side of *ortho*-fluorine substitution (entry 8). Note that in entries 9 and 10 some di- or triarylated by-products were observed. For entry 9, the yields for mono-, di-, and triarylated products were 14, 21, and 49%, respectively. For entry 10, the yields for mono- and diarylated products were 31 and 58%. These results revealed that the direct arylation of acidic C–H bonds of polyfluoroarenes [11, 12] is a side reaction in Pd-catalyzed decarboxylative coupling of polyfluorobenzoates.

4.1.3 Mechanistic Study

To understand the mechanism of the Pd-catalyzed decarboxylative cross-coupling, we carried out DFT calculations [14] (Fig. 4.1). Through the analysis, we first concluded that it is Pd(II), but not Pd(0) which does not have the ability to catalyze decarboxylation of carboxylic acid, that promotes the decarboxylation of the carboxylic acid. Accordingly, we proposed that in a catalytic cycle a monocoordinated Pd(0) complex first undergoes oxidative addition with the aryl halide [15, 16]. When PMe$_3$ is used as a model ligand, the Pd(0) complex forms a complex with PhCl (**IN1**), which undergoes oxidative addition to produce a Pd(II) intermediate (**IN2**) through **TS1**. The energy barrier for oxidative addition is +14.5 kcal/mol. **IN2** then exchanges the anionic ligand to form **IN3** and its coordination isomer **IN4**. Note that the barrier between **IN3** and **IN4** is only +16.8 kcal/mol meaning

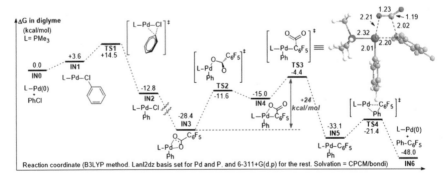

Fig. 4.1 Proposed mechanism for Pd-catalyzed decarboxylative cross-coupling. Reprinted with the permission from Org. Lett. **2010**, 12, 1000. Copyright 2011 American Chemical Society

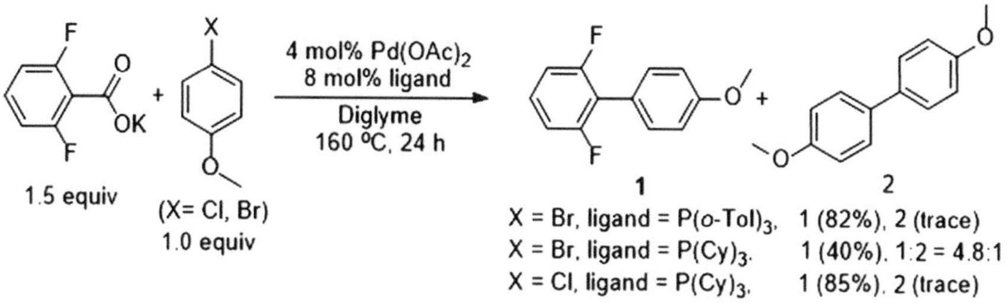

Scheme 4.1 Effects of the ligands on the cross-coupling. Reprinted with the permission from Org. Lett. 2010, 12, 1000. Copyright 2011 American Chemical Society

that they can readily isomerize each other. From **IN4**, the decarboxylation transition state (**TS3**) is identified as a four-coordinate Pd(II) species. From **IN3** to **TS3**, the free energy increases by 24.0 kcal/mol, and therefore, decarboxylation is the rate-limiting step. The immediate product of decarboxylation is a three-coordinate Pd(II) complex (**IN5**), which undergoes reductive elimination to form the biaryl product through **TS4** with a low barrier of 11.7 kcal/mol.

The above results reveal a major difference between Cu- and Pd-catalyzed decarboxylative couplings. In Cu-catalyzed versions in previous chapter, the rate-limiting step is oxidative addition of the Cu(I) intermediate to aryl halide [8]. This conclusion is consistent with the observation that aryl chlorides cannot be converted by the copper catalyst. By comparison, in Pd-catalyzed versions the rate-limiting step should be decarboxylation. To support this mechanistic proposal, we compared the Pd-catalyzed decarboxylative couplings of potassium 2,6-difluorobenzoate with 4-methoxyphenyl bromide and chloride (Scheme 4.1). For the bromide, the more electron-rich ligand (PCy$_3$) affords a worse result due to the side reaction, whereas the less electron-rich ligand P(o-Tol)$_3$ does not show this problem. The explanation is presumably that the more electron-rich ligand causes a faster oxidative addition and therefore produces the aryl-Pd(II) intermediate too rapidly to be effectively consumed by the benzoate. On the other hand, for the aryl chloride that will undergo oxidative addition less rapidly, the use of the PCy$_3$ ligand does not cause much side reaction. Thus, a good balance between the decarboxylation and oxidative addition steps is critical to the success of decarboxylative cross-couplings. Goosen et al. use a bimetallic catalyst (e.g., Pd for oxidative addition and Cu for decarboxylation) to control the balance [3]. We show here that under some circumstances a strategic change of ligand can also control the balance.

4.1.4 The Cross-Coupling Controlled by the Ligands

Note that the ligand effects can be used for selective decarboxylative cross-couplings. A good example is shown in Scheme 4.2, two different

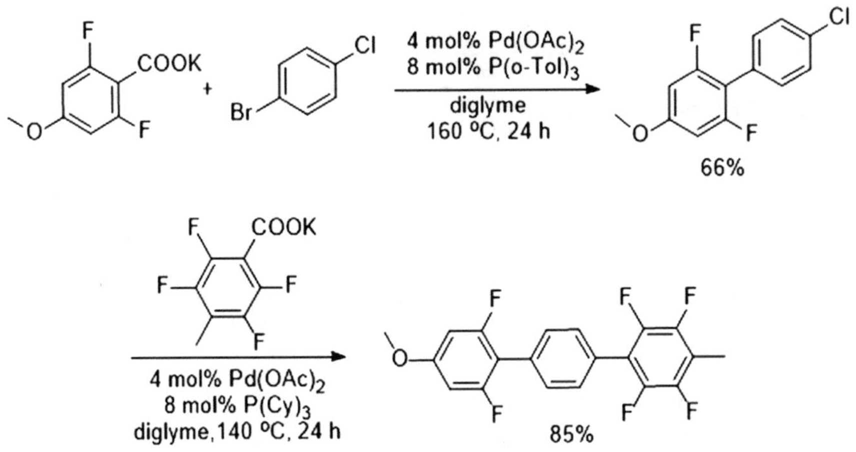

Scheme 4.2 Selective decarboxylative cross-couplings. Reprinted with the permission from Org. Lett. 2010, 12, 1000. Copyright 2011 American Chemical Society

polyfluorophenyl groups are sequentially coupled to an aryl halide containing both aryl bromides and aryl chlorides bond.

4.2 Conclusion

In summary, we report Pd-catalyzed decarboxylative cross-coupling of potassium polyfluorobenzoates with aryl bromides, chlorides, and triflates. The reaction is practical for synthesis of polyfluorobiaryls from readily accessible and nonvolatile polyfluorobenzoate salts. Unlike the Cu-catalyzed decarboxylation cross-coupling where oxidative addition is the rate-limiting step, in the Pd-catalyzed version decarboxylation is the rate-limiting step.

4.3 Experimental Section and Compound Data

4.3.1 General Information

All the reactions were carried out in oven-dried Schlenk tubes under Argon atmosphere (purity ≥ 99.999%). The diglyme solvent was bought from Sigma-Aldrich (Sure/Seal) without further purification. All aryl halides were bought from Alfa Aesar or Acros and used as is. Aryl triflates and tosylates were synthesized according to the literature methods. All the polyfiuorobenzoic acids were purchased from Sigma-Aldrich. Palladium(II) acetate was purchased from Alfa Aesar (99.98% purity). All phosphine ligands were bought from Sigma-Aldrich,

Strem or Alfa Aesar and used as is. All the other reagents and solvents mentioned in this text were bought from Sinopharm Chemical Reagent Co. Ltd or Alfa Aesar and purified when necessary.

^{1}H-NMR, ^{13}C-NMR, ^{19}F-NMR spectra were recorded on a Bruker Avance 400 spectrometer at ambient temperature in $CDCl_3$ unless otherwise noted. Data for ^{1}H-NMR are reported as follows: chemical shift (δ m), multiplicity, integration, and coupling constant (Hz). Data for ^{13}C-NMR are reported in terms of chemical shift (δ ppm), multiplicity, and coupling constant (Hz). ^{19}F-NMR was recorded in $CDCl_3$ solutions and trifluorotoluene (TFT) ($\delta = -67.73$ ppm) was employed as an external standard. Data for ^{19}F-NMR are reported as follows: chemical shift (δ m), multiplicity, integration, and coupling constant (Hz).

Gas chromatographic (GC) analysis was acquired on a Shimadzu GC-2014 Series GC System equipped with a flame ionization detector. GC-MS analysis was performed on Thermo Scientific AS 3000 Series GC-MS System. HRMS analysis was performed on Finnigan LCQ advantage Max Series MS System. Elementary Analysis was carried out on Elementar Vario EL III elemental analyzer. Organic solutions were concentrated under reduced pressure on a Buchi rotary evaporator. Flash column chromatographic purification of products was accomplished using forced-flow chromatography on Silica Gel (200–300 mesh).

4.3.2 General Procedures

The coupling of potassium pentafluorobenzoate with aryl chlorides, bromides, and triflates:

A 10 ml oven-dried Schlenk tube was charged with $Pd(OAc)_2$ (1–2% mol), ligand (2–4% mol PCy_3 or $P(o\text{-Tol})_3$), appointed amount of potassium pentafluorobenzoate (0.30–0.375 mmol) and the corresponding aryl electrophile (0.25 mmol) (if solid). The tube was evacuated and filled with argon (this procedure was repeated three times). Then aryl electrophile (0.25 mmol) (if liquid) and diglyme (0.5 ml) were added with a syringe under a counter flow of argon. The tube was sealed with a screw cap, stirred at room temperature for 10 min, and connected to the Schlenk line which was full of argon, stirred in a preheated oil bath (130 °C) for the appointed time (24 h). Upon completion of the reaction, the mixture was cooled to room temperature, diluted with diethyl ether (20 ml), and filtered through a short silica gel column to remove the deposition. The organic layers were washed with water (20 ml × 3), and then with brine, dried over Na_2SO_4, and filtered, the solvents were removed via rotary vapor. Purification of the residue by column chromatography (silica gel, ethyl acetate/hexane gradient) yielded the corresponding fluoroarene.

4.3.3   Characterization of the Products

Compound name 2,3,4,5,6-pentafluorobiphenyl

54 mg (89%) of a white solid was obtained. This compound is known. Reference: Korenaga, T.; Kosaki, T.; Fukumura, R.; Ema, T.; Sakai, T. Org. Lett. 2005, 7, 4915–4918. ^{1}H-NMR (400 MHz, CDCl$_3$, 293 K, TMS): δ 7.39–7.42 (m, 2H), 7.44–7.48 (m, 3H). ^{13}C-NMR (100 MHz, CDCl$_3$, 293 K, TMS): δ 115.9 (td, J_F = 3.8, 17.5 Hz), 126.4, 128.7, 129.3, 130.1, 137.8 (dm, J_F = 253.2 Hz), 140.4 (dm, J_F = 253.2 Hz), 144.2 (dm, J_F = 252.4 Hz). ^{19}F-NMR (377 MHz, CDCl$_3$, 293 K, TFT): −162.3 (m, 2F), −155.6 (t, J_F = 21.1 Hz, 1F), −143.2 (dd, J_F = 23.0, 8.3 Hz, 2F).

Compound name 2,3,4,5,6-pentafluoro-2'-methoxybiphenyl

64 mg (94%) of a white solid was obtained. This compound is known. Reference: Lafrance, M.; Shore, D.; Fagnou, K. Org. Lett. 2006, 8, 5097–5100. ^{1}H-NMR (400 MHz, CDCl$_3$, 293 K, TMS): δ 3.80 (s, 3H), 7.01–7.07 (m, 2H), 7.21–7.23 (m, 1H), 7.43–7.46 (m, 1H). ^{13}C-NMR (100 MHz, CDCl$_3$, 293 K, TMS): δ 55.6, 111.3, 112.8 (t, J_F = 19.4 Hz), 115.3, 120.6, 131.1, 131.7, 137.6 (dm, J_F = 249.3 Hz), 140.5 (dm, J_F = 251.1 Hz), 144.5 (dm, J_F = 248.3 Hz), 157.2. ^{19}F-NMR (377 MHz, CDCl$_3$, 293 K, TFT): −163.2 (m, 2F), −156.2 (t, J_F = 20.7 Hz, 1F), −140.3 (dd, J_F = 23.0, 8.3 Hz, 2F).

Compound name 2,3,4,5,6-pentafluoro-3'-methoxybiphenyl

65 mg (96%) of a white solid was obtained. This compound is known. Reference: Lafrance, M.; Rowley, C. N.; Woo, T. K.; Fagnou, K. J. Am. Chem. Soc. 2006, 128, 8754–8756. ^{1}H-NMR (400 MHz, CDCl$_3$, 293 K, TMS): δ 3.84 (s, 3H), 6.94 (s, 1H), 6.98–7.01 (m, 2H), 7.38–7.42 (t, J = 8.0 Hz, 1H). ^{13}C-NMR (100 MHz, CDCl$_3$, 293 K, TMS): δ 55.2, 114.8 (td, J_F = 3.4, 17.2 Hz), 114.8, 115.8, 122.4, 127.5, 129.7, 137.8 (dm, J_F = 250.8 Hz), 140.4 (dm, J_F = 252.0 Hz), 144.2 (dm, J_F = 246.4 Hz), 159.7. ^{19}F-NMR (377 MHz, CDCl$_3$, 293 K, TFT): −162.2 (m, 2F), −155.6 (t, J_F = 20.7 Hz, 1F), −142.8 (dd, J_F = 23.0, 8.3 Hz, 2F).

Compound name 5-(perfluorophenyl)benzo[d][1,3] dioxole

63 mg (87%) of a white solid was obtained. This compound is known. Reference: Shang, R.; Fu, Y.; Wang, Y.; Xu, Q.; Yu, H. Z.; Liu, L. Angew. Chem. 2009, 121, 9514–9518; Angew. Chem. Int. Ed. 2009, 48, 9350–9354. ^{1}H-NMR (400 MHz, $CDCl_3$, 293K, TMS): δ 6.04 (s, 2H), 6.88–6.95 (m, 3H). ^{13}C-NMR (100 MHz, $CDCl_3$, 293 K, TMS): δ 101.5, 108.6, 110.3, 115.7 (td, J_F = 3.8, 17.2 Hz), 119.4, 124.2, 137.8 (dm, J_F = 250.8 Hz), 140.2 (dm, J_F = 252.0 Hz), 144.2 (dm, J_F = 245.6 Hz), 148.0, 148.5. ^{19}F-NMR (377 MHz, $CDCl_3$, 293K, TFT): −162.3 (m, 2F), −156.0 (t, J_F = 20.7 Hz, 1F), −143.2 (dd, J_F = 23.4, 8.3 Hz, 2F). HRMS calcd for $C_{13}H_5F_5O_2$ (M+) 288.0210; found: 288.0212.

Compound name 1-(perfluorophenyl) naphthalene

71 mg (97%) of a white solid was obtained. This compound is known. Reference: Do, H. Q.; Daugulis, O. J. Am. Chem. Soc. 2008, 130, 1128–1129; ^{1}H-NMR (400 MHz, $CDCl_3$, 293 K, TMS): δ 7.45–7.63 (m, 5H), 7.94–8.04 (m, 2H). ^{13}C-NMR (100 MHz, $CDCl_3$, 293K, TMS): δ 144.4 (td, J_F = 3.1, 19.5 Hz), 123.8, 124.5, 125.2, 126.4, 127.1, 128.6, 129.0, 130.1, 131.5, 133.7, 137.8 (dm, J_F = 251.7 Hz), 140.9 (dm, J_F = 252.5 Hz), 144.6 (dm, J_F = 246.1 Hz). ^{19}F-NMR (377 MHz, $CDCl_3$, 293 K, TFT): −161.9 (m, 2F), −154.7 (t, J_F = 20.7 Hz, 1F), −139.4 (dd, J_F = 22.6, 7.9 Hz, 2F).

Compound name 2,3,4,5,6-pentafluoro-4′-nitrobiphenyl

70 mg (97%) of a pale yellow solid was obtained. This compound is known. Reference: Korenaga, T.; Kosaki, T.; Fukumura, R.; Ema, T.; Sakai, T. Org. Lett. 2005, 7, 4915–4918. ^{1}H-NMR (400 MHz, $CDCl_3$, 293K, TMS): δ 7.64–7.66 (m, 2H), 8.35–8.38 (m, 2H). ^{13}C-NMR (100 MHz, $CDCl_3$, 293 K, TMS): δ 113.8 (td, J_F = 3.4, 17.2 Hz), 123.8, 131.3, 132.9, 138.0 (dm, J_F = 252.4 Hz), 141.3 (dm, J_F = 254.8 Hz), 144.0(dm, J_F = 259.0 Hz), 148.2. ^{19}F-NMR (377 MHz, $CDCl_3$, 293 K, TFT): −160.8 (m, 2F), −152.5 (t, J_F = 20.7 Hz, 1F), −142.5 (dd, J_F = 22.6, 8.7 Hz, 2F).

Compound name 2′,3′,4′,5′,6′-pentafluorobiphenyl-4-carbonitrile

66 mg (98%) of a pale yellow solid was obtained. This compound is known. Reference: Lafrance, M.; Shore, D.; Fagnou, K. Org. Lett. 2006, 8, 5097–5100.

^{1}H-NMR (400 MHz, CDCl₃, 293 K, TMS): δ 7.55–7.57 (m, 2H), 7.79–7.81 (m, 2H). ^{13}C-NMR (100 MHz, CDCl₃, 293K, TMS): δ 113.3, 114.1 (td, J_F = 3.1, 17.7 Hz), 118.0, 130.9, 131.1, 132.4, 137.9 (dm, J_F = 251.8 Hz), 141.1 (dm, J_F = 254.9 Hz), 144.0 (dm, J_F = 247.4 Hz). ^{19}F-NMR (377 MHz, CDCl₃, 293K, TFT): −160.9 (m, 2F), −152.8 (t, J_F = 21.5 Hz, 1F), −142.7 (dd, J_F = 22.6, 7.2 Hz, 2F).

Compound name 2-fluoro-4′-methoxy-6-(trifluoromethyl) biphenyl

50 mg (74%) of a colorless liquid was obtained. This compound is known. Reference: Shang, R.; Fu, Y.; Wang, Y.; Xu, Q.; Yu, H. Z.; Liu, L. Angew. Chem. 2009, 121, 9514–9518; Angew. Chem. Int. Ed. 2009, 48, 9350–9354. ^{1}H-NMR (400 MHz, CDCl₃, 293K, TMS): δ 3.86 (s, 3H), 6.95–6.98 (dm, J = 8.8 Hz, 2H), 7.21 (d, J = 8.8 Hz, 2H), 7.31 (t, J = 8.8 Hz, 1H), 7.40–7.46 (m, 1H), 7.55 (d, J = 7.6 Hz, 1H). ^{13}C-NMR (100 MHz, CDCl₃, 293 K, TMS): δ 55.2, 113.4, 119.0 (d, J_F = 23.5 Hz), 121.6 (m), 123.4 (qd, J_F = 3.6, 272.7 Hz), 123.8, 128.8 (d, J_F = 8.7 Hz), 129.0 (d, J_F = 16.0 Hz), 130.9, 131.1 (qd, J_F = 2.7, 29.4 Hz), 159.6, 160.5 (d, J_F = 243.8 Hz). ^{19}F-NMR (377 MHz, CDCl₃, 293 K, TFT): −111.7 (s, 1F), −57.3 (s, 3F). HRMS calcd for $C_{14}H_{10}F_4O$ (M+) 270.0668; found: 270.0672.

Compound name 2,3,6-trifluoro-4′-methoxybiphenyl

30 mg (51%) of a white solid was obtained. This compound is known. Reference: Shang, R.; Fu, Y.; Wang, Y.; Xu, Q.; Yu, H. Z.; Liu, L. Angew. Chem. 2009, 121, 9514–9518; Angew. Chem. Int. Ed. 2009, 48, 9350–9354. ^{1}H-NMR (400 MHz, CDCl₃, 293K, TMS): δ 3.86 (s, 3H), 6.87–6.93 (m, 1H), 6.99–7.02(m, 2H), 7.04–7.12 (m, 1H), 7.39–7.42 (m, 2H). ^{13}C-NMR (100 MHz, CDCl₃, 293 K, TMS): δ 55.3, 110.7 (dm, J_F = 25.5 Hz), 113.9, 115.1 (qd, J_F = 1.1, 9.4 Hz), 120.0 (dd, J_F = 14.9, 20.4 Hz), 120.4 (d, J_F = 2.0 Hz), 131.4, 147.45 (dd, J_F = 3.6, 242.9 Hz), 147.9 (dq, J_F = 7.4, 247.9 Hz), 155.3 (dm, J_F = 242.7 Hz), 159.8. ^{19}F-NMR (377 MHz, CDCl₃, 293 K, TFT): −142.2 (dd, J_F = 15.1, 21.9 Hz, 1F),

-138.2 (dd, $J_F = 3.8$, 21.5 Hz, 1F), -120.0 (dd, $J_F = 4.1$, 15.1 Hz, 1F). HRMS calcd for $C_{13}H_9F_3O$ (M+) 238.0605; found: 238.0600.

Compound name 2,3,5,6-tetrafluoro-4'-methoxy-4-methylbiphenyl

65 mg (96%) of a white solid was obtained. This compound is known. Reference: Shang, R.; Fu, Y.; Wang, Y.; Xu, Q.; Yu, H. Z.; Liu, L. Angew. Chem. 2009, *121*, 9514–9518; Angew. Chem. Int. Ed. 2009, 48, 9350–9354. ^{1}H-NMR (400 MHz, CDCl$_3$, 293K, TMS): δ 2.31 (t, $J_F = 2.0$ Hz, 3H), 3.86 (s, 3H), 7.00 (d, $J = 8.8$ Hz, 2H), 7.39 (d, $J = 8.8$ Hz, 2H). ^{13}C-NMR (100 MHz, CDCl$_3$, 293 K, TMS): δ 7.44, 55.2, 114.0, 114.4 (t, $J_F = 19.1$ Hz), 117.6 (t, $J_F = 16.5$ Hz), 119.8, 131.4, 143.7 (dm, $J_F = 243.3$ Hz), 145.3 (dm, $J_F = 242.3$ Hz), 160.0. ^{19}F-NMR (377 MHz, CDCl$_3$, 293K, TFT): -146.0 (dd, $J_F = 12.4$, 22.2 Hz, 2F), -144.4 (dd, $J_F = 12.8$, 22.6 Hz, 2F). HRMS calcd for $C_{14}H_{10}F_4O$ (M+) 270.0668; found: 270.0666.

Compound name 2,3,5,6-tetrafluoro-4'-methoxybiphenyl-4-amine

58 mg (85%) of a yellow solid was obtained. This compound is new. ^{1}H-NMR (400 MHz, CDCl$_3$, 293K, TMS): δ 3.85 (s, 3H), 3.99 (s, br, 2H), 6.98 (d, $J = 8.8$ Hz, 2H), 7.35 (d, $J = 8.8$ Hz, 2H), ^{13}C-NMR (100 MHz, CDCl$_3$, 293K, TMS): δ 55.3, 108.1 (t, $J_F = 17.1$ Hz), 114.0, 120.1, 124.7–125.0 (m), 131.5, 136.9 (dm, $J_F = 236.6$ Hz), 144.1 (dm, $J_F = 240.5$ Hz), 159.5. ^{19}F-NMR (377 MHz, CDCl$_3$, 293K, TFT): -162.4 (m, 2F), -146.9 (m, 2F). HRMS calcd for $C_{14}H_{10}F_4O$ (M+) 271.0620; found: 271.0627.

Compound name 1,3-di-(*p*-methoxyphenyl)-2,4,6-trifluorobenzene

18 mg (21%) of a white solid was obtained. This compound is known. Reference: Shang, R.; Fu, Y.; Wang, Y.; Xu, Q.; Yu, H. Z.; Liu, L. Angew. Chem. 2009, 121, 9514–9518; Angew. Chem. Int. Ed. 2009, 48, 9350–9354. ^{1}H-NMR (400 MHz, CDCl$_3$, 293K, TMS): δ 3.85 (s, 6H), 6.84 (td, $J_F = 2.4$, 9.6 Hz, 1H), 7.00 (d, $J = 8.8$ Hz, 4H), 7.38 (d, $J = 8.4$ Hz, 4H). ^{13}C-NMR (100 MHz, CDCl$_3$, 293K, TMS): δ 55.3, 100.2 (td, $J_F = 4.4$, 15.5 Hz), 113.9, 114.8 (m), 120.8, 131.5, 158.5 (dm, $J_F = 246.4$ Hz), 158.7 (dm, $J_F = 246.4$ Hz), 159.5. ^{19}F-NMR (377 MHz, CDCl$_3$, 293K, TFT): -115.4 (t, $J_F = 6.8$ Hz, 1F), -113.8 (d, $J_F = 6.4$ Hz, 2F). HRMS calcd for $C_{20}H_{15}F_3O_2$ (M+) 344.1024; found: 344.1023.

References

1. (a) Baudoin, O. (2007). *Angewandte Chemie International Edition, 46*, 1373–1375; (b) Goossen, L. J., Rodriguez, N., & Goossen, K. (2008). *Angewandte Chemie International Edition, 47*, 3100–3120.

2. (a) Goossen, L. J., Deng, G., & Levy, L. M. (2006). *Science, 313*, 662–664. (b) Goossen, L. J., Rodriguez, N., Melzer, B., Linder, C., Deng, G., & Levy, L. M. (2007). *Journal of the American Chemical Society, 129*, 4824–4833. (c) Goossen, L. J., & Melzer, B. (2007). *The Journal of Organic Chemistry, 72*, 7473–7476. (d) Goossen, L. J., & Knauber, T. (2008). *The Journal of Organic Chemistry, 73*, 8631–8634. (e) Goossen, L. J., Zimmermann, B., & Knauber, T. (2008). *Angewandte Chemie International Edition, 47*, 7103–7106. (f) Goossen, L. J., Rodriguez, N., & Linder, C. (2008). *Journal of the American Chemical Society, 130*, 15248–15249. (g) Goossen, L. J., Rudolphi, F., Oppel, C., & Rodriguez, N. (2008). *Angewandte Chemie International Edition, 47*, 3043–3045.

3. (a) Myers, A. G., Tanaka, D., & Mannion, M. R. (2009). *Journal of the American Chemical Society*, 11250–11251. (b) Tanaka, D., & Myers, A. G. (2004). *Organic Letters, 6*, 433–436. (c) Tanaka, D., Romeril, S. P., & Myers, A. G. (2005). *Journal of the American Chemical Society, 127*, 10323–10333.

4. Forgione, P., Brochu, M.-C., St-Onge, M., Thesen, K. H., Bailey, M. D., & Bilodeau, F. (2006). *Journal of the American Chemical Society, 128*, 11350–11351.

5. (a) Moon, J., Jeong, M., Nam, H., Ju, J., Moon, J. H., Jung, H. M., & Lee, S. (2008). *Organic Letters, 10*, 945–948. (b) Shang, R., Fu, Y., Li, J. B., Zhang, S. L., Guo, Q. X., & Liu, L. (2009). *Journal of the American Chemical Society, 131*, 5738–5739.

6. (a) Waetzig, S. R., Rayabarupu, D. K., Weaver, J. D., & Tunge, J. A. (2006). *Angewandte Chemie International Edition, 45*, 4977–4980. (b) Waetzig, S. R., & Tunge, J. A. (2007). *Journal of the American Chemical Society, 129*, 14860–14861.

7. Shang, R., Fu, Y., Wang, Y., Xu, Q., Yu, H.-Z., & Liu, L. (2009). *Angewandte Chemie International Edition, 48*, 9350–9354.

8. Sakai *et al.* developed a special protocol for the Suzuki coupling reaction of pentafluorophenylboronic acid, which is an inactive substrate under normal conditions. See: Korenaga, T., Kosaki, T., Fukumura, R., Ema, T., & Sakai, T. (2005). *Organic Letters, 7*, 4915–4917.

9. (a) Nitschke, J. R., & Tilley, T. D. (2001). *Journal of the American Chemical Society, 123*, 10183–10190; (b) Zacharias, P., Gather, M. C., Rojahn, M., Nuyken, O., & Meerholz, K. (2007). *Angewandte Chemie International Edition, 46*, 4388–4392.

10. (a) Mewshaw, R. E., Edsall Jr., R. J., Yang, C., Manas, E. S., Xu, Z. B., Henderson, R. A., Keith Jr., J. C., & Harris, H. A. (2005). Journal of Medicinal Chemistry, 48, 3953–3979. (b) de Candia, M., Liantonio, F., Carotti, A., De Cristofaro, R., & Altomare, C. (2009). Journal of Medicinal Chemistry, 52, 1018–1028.

11. (a) Lafrance, M., Rowley, C. N., Woo, T. K., & Fagnou, K. (2006). *Journal of the American Chemical Society, 128*, 8754–8756. (b) Lafrance, M., Shore, D., & Fagnou, K. (2006). *Organic Letters, 8*, 5097–5100.

12. (a) Do, H.-Q., & Daugulis, O. (2007). *Journal of the American Chemical Society, 129*, 12404–12405. (b) Do, H.-Q., & Daugulis, O. (2008). *Journal of the American Chemical Society, 130*, 1128–1129. (c) Do, H.-Q., Kashif Khan, R. M., & Daugulis, O. (2008). *Journal of the American Chemical Society, 130*, 15185–15192.

13. Surry, D. S., & Buchwald, S. L. (2008). *Angewandte Chemie International Edition, 47*, 6338–6361.

14. For the previous theoretical analysis on the related subject, see: (a) Goossen, L. J., Thiel, W. R., Rodriguez, N., Linder, C., & Melzer, B. (2007). *Advanced Synthesis & Catalysis,, 349*, 2241–2246. (b) Zhang, S.-L., Fu, Y., Shang, R., Guo, Q.-X., & Liu, L. (2010). *Journal of the American Chemical Society, 132*, 638–646.

15. Oxidative addition is not a rate-limiting step in the decarboxylative cross coupling because a mechanistic related Pd-catalyzed cross coupling reaction (e.g. Suzuki coupling) does not require a high temperature to proceed. Therefore, we do not discuss the detailed mechanism of the oxidative addition step, which may be complicated by many factors. For oxidative addition of an aryl halide to monoligated Pd(0), see: (a) Ahlquist, M., Fristrup, P., Tanner, D., & Norrby, P.-O. (2006). *Organometallics, 26*, 2066–2073. (b) Li, Z., Fu, Y., Guo, Q.-X., & Liu, L. (2008). *Organometallics, 27*, 4043–4049.
16. Amatore, C., Jutand, A., & Suarez, A. (1993). *Journal of the American Chemical Society, 115*, 9531–9541.

Chapter 5
Construction of C(sp³)–C(sp²) Bonds Via Palladium-Catalyzed Decarboxylative Couplings of 2-(2-Azaaryl)Acetate Salts with Aryl Halides

Abstract Pd-catalyzed decarboxylative cross-couplings of 2-(2-azaaryl)acetates with aryl halides and triflates have been discovered. This reaction is potentially useful for the synthesis of some functionalized pyridines, quinolines, pyrazines, benzoxazoles, and benzothiazoles. Theoretical analysis shows that the nitrogen atom at the 2-position of the heteroaromatics directly coordinates to Pd(II) in the decarboxylation transition state.

5.1 Introduction

Transition metal-catalyzed decarboxylative cross-coupling has received widespread attention in synthetic chemistry because the method uses carboxylic acids as alternative reagents over organometallic compounds [1]. Myers [2], Forgione [3], and others [4–7], including what we discussed in previous chapters [8] found that Pd could catalyze the decarboxylative coupling of aromatic, alkenyl, and alkynyl carboxylic acids. Goossen et al. developed excellent procedures for the Pd/Cu-catalyzed decarboxylative coupling of benzoic acids and alfa-oxocarboxylates [9]. Our group described the Cu-catalyzed decarboxylative coupling of polyfluorobenzoic acids [10] (In Chap. 3). Related decarboxylative reactions of some aliphatic esters were described recently by Tunge, Trost, Stoltz, and others [11, 12]. Here we report a new example for breaking the C_{sp^3}–COOH bond: the Pd-catalyzed

© Springer Nature Singapore Pte Ltd. 2017

R. Shang, *New Carbon–Carbon Coupling Reactions Based on Decarboxylation and Iron-Catalyzed C–H Activation*, Springer Theses, DOI 10.1007/978-981-10-3193-9_5

decarboxylative couplings of 2-(2-azaaryl)acetates with aryl halides and triflates. This study was inspired by the recent fascinating work of Oshima et al. on chelation-assisted activation of CC_{sp^3}–CC_{sp^3} bonds [13].

5.2   Results and Discussion

5.2.1   Investigation of the Reaction Conditions

Our study began by testing the Pd-catalyzed decarboxylative coupling of potassium 2-(2-pyridyl)acetate with bromobenzene (Table 5.1). We examined a series of Pd salts and phosphine ligands. We found the Xant-Phos was the best ligand in this reaction (entry 12). The desired product was obtained in 96% yield using Xant-Phos. Chlorobenzene could not be efficiently converted under the same conditions (entry 16), whereas phenyl triflate affords the desired product in 84% yield (entry 17). It is interesting to note that the related potassium salts of 2-(3-pyridyl)-, 2-(4-pyridyl)-, and 2-phenyl acetic acids could not undergo this coupling reaction [14]. It is also

Table 5.1 Decarboxylative coupling under various conditions[a]

Entry	X	Pd source	Ligand	Solvent	Yield[a] (%)
1	Br	Pd(OAc)$_2$	P(Cy)$_3$	Diglyme	28
2	Br	Pd(OAc)$_2$	S-Phos	Diglyme	37
3	Br	Pd(OAc)$_2$	Dave-Phos	Diglyme	36
4	Br	Pd(OAc)$_2$	X-Phos	Diglyme	29
5	Br	Pd(OAc)$_2$	S-BINAP	Diglyme	68
6	Br	Pd(OAc)$_2$	Tol-BINAP	Diglzyme	65
7	Br	Pd(OAc)$_2$	DPPP	Diglyme	4
8	Br	Pd(OAc)$_2$	Xant-Phos	Diglyme	74
9	Br	Pd(OAc)$_2$	DPE-Phos	Diglyme	72
10	Br	Pd(OAc)$_2$	DPPF	Diglyme	26
11	Br	Pd(acac)$_2$	Xant-Phos	Diglyme	37
12[b]	Br	Pd$_2$(dba)$_3$	Xant-Phos	Diglyme	96 (96[c])
13	Br	Pd(TFA)$_2$	Xant-Phos	Diglyme	48
14	Br	Pd(MeCN)$_2$Cl$_2$	Xant-Phos	Diglyme	53
15[b]	Br	[PdCl(allyl)]$_2$	Xant-Phos	Diglyme	82
16[b]	Cl	Pd$_2$(dba)$_3$	Xant-Phos	Diglyme	Trace
17[b]	OTf	Pd$_2$(dba)$_3$	Xant-Phos	Mesitylene	87 (84[c])

[a]GC yields (average of two runs) with naphthalene as internal standard. All the reactions were carried out at 0.25 mmol scale in 0.5 mL solvent with 6 mol% bidentate ligand or 8 mol% monodentate ligand. [b]Use of 2 mol% Pd salt. [c]Isolated yields

important to point out that the intramolecular decarboxylative coupling of cinnamyl 2-(2-pyridyl)acetate esters was reported previously by Waetzig and Tunge [11b].

5.2.2   Exploration of the Substrate Scope

With the optimized conditions in hand, we explored the scope of this reaction with various aryl halides and triflates (Table 5.2). We found that both electron-rich and electron-deficient aryl bromides can be successfully converted across a range of functional groups. The yields of aryl triflates are similar to those for aryl bromides.

Table 5.2 Decarboxylative coupling of potassium 2-(2-Pyridyl)acetate[a]

Reprinted with the permission from J. Am. Chem. Soc., 2010, 132, 14391. Copyright 2011 American Chemical Society

[a]Isolated yields. All the reactions were carried out at 0.30 mmol scale in 0.6 mL solvents

Table 5.3 Decarboxylative cross-coupling of 2-(2-Pyridyl)acetic acid derivatives[a]

3bz, Trace 3cz, 62%

R = H   3dz, 92%
R = OMe 3dc, 71%
R = CF$_3$ 3dw, 67%

3ez, 91%

3fz, 78% in mesitylene 3gz, 86%

3hz, 65% in diglyme
70 % in mesitylene

3iz, R = H   95%
3ib, R = 4-F 80%
3il, R = 3,5-di-Me 84%
Solvent = mesitylene

Reprinted with the permission from J. Am. Chem. Soc., 2010, 132, 14391. Copyright 2011 American Chemical Society

[a]Isolated yields. All the reactions were carried out at 0.30 mmol scale in 0.6 mL solvent

An activated aryl chloride (in the case of **3aw**) may also be used in the reaction. The yield of product is 66%.

Moreover, many derivatives of 2-(2-pyridyl)acetate can be successfully converted (Tables 5.3 and 5.4) to functionalized pyridines, quinolines, pyrazines, benzoxazoles, and benzothiazoles. Recently Fagnou's work showed that benzylic arylation of pyridine derivatives could be accomplished through picoline N-oxide sp^3 arylation followed by deoxygenation [15]. The decarboxylative cross-coupling described here provides an alternative approach to synthesize related heterocyclic compounds that may be useful in medicinal chemistry.

5.2.3   Mechanistic Study

To understand how the pyridyl group assisted the decarboxylative coupling, we conducted standard DFT calculations [16] (Fig. 5.1). First of all, a bis-ligated Pd(0) complex is proposed to activate the aryl halide to produce a Pd(II) intermediate (**CP2**) through **TS1** [17]. The energy barrier for oxidative addition is +21.2 kcal/mol. **CP2** then exchanges the anion to form **CP3** and its isomer **CP4**. From **CP4**, the decarboxylation transition state (**TS2**) was identified as a four-coordinate Pd(II) species. In **TS2**, the Pd(II) coordinates to the pyridyl nitrogen, but not to the leaving CO$_2$ moiety. From **CP2** to **TS2** the free energy increases by +33.9 kcal/mol, a value consistent with the temperature of the reaction (Note: a first-order reaction with a half-life of 12 h at 150 °C should have a barrier of

Table 5.4 Decarboxylative cross-coupling of benzooxazol-2-yl-acetate and Benzothiazol-2-yl-acetate[a]

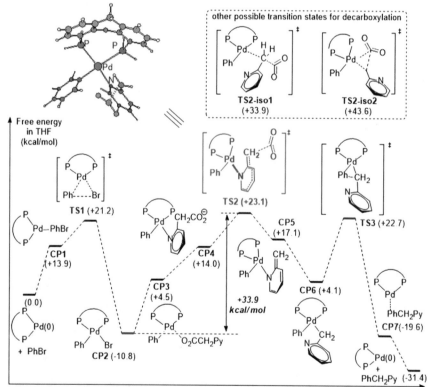

Reprinted with the permission from J. Am. Chem. Soc., 2010, 132, 14391. Copyright 2011 American Chemical Society

[a]Isolated yields. All the reactions were carried out at 0.30 mmol scale in 1.0 mL solvents

Fig. 5.1 Proposed mechanism. Reprinted with the permission from J. Am. Chem. Soc., 2010, 132, 14391. Copyright 2011 American Chemical Society

+34.3 kcal/mol). Subsequent reductive elimination through **TS3** affords the desired product. In the catalytic cycle, the decarboxylation step is rate-limiting. The coordination of nitrogen to Pd(II) is crucial to reducing the energy barrier of decarboxylation, which may be attributed to a stabilized anion effect analogous to the enolate chemistry described by Tunge and others [11, 18]. Also, much higher energies were calculated for the other two possible transition states (i.e., **TS2-iso1** in which Pd(II) interacts with the α-carbon atom, and **TS2-iso2** in which Pd(II) interacts with both the α-carbon atom and CO$_2$ moiety). Furthermore, it is important to note that a transition state corresponding to decarboxylation through chelation of Pd with O and N atoms in the substrate could not be found.

5.3 Conclusion

In summary, Pd-catalyzed decarboxylative cross-coupling of 2-(2-azaaryl)acetates with aryl halides and triflates is discovered. This reaction is potentially useful for the synthesis of various functionalized azaarenes. Theoretical analysis indicates that the nitrogen atom at the 2-position of the heteroaromatics directly coordinates to Pd(II) in the decarboxylation transition state.

5.4 Experimental Section and Compound Data

5.4.1 General Information

All the reactions were carried out in oven-dried Schlenk tubes under Argon atmosphere (purity ≥ 99.999%). The diglyme solvent was bought from Sigma-Aldrich (Sure/Seal) and used without further purification. The mesitylene solvent was bought from Sigma-Aldrich and used without further purification. All aryl halides were bought from Alfa Aesar or Acros and used as is. Aryl triflates (Ref. J. Org. Chem. 2004, 69, 1137) and tosylates (Ref. J. Am. Chem. Soc., 2008, 130, 2754.) were synthesized according to the literature methods. Potassium 2-azaaryl acetates were synthesized by saponification of the corresponding methyl ester. The methyl 2-azaaryl acetates were synthesized according to the literature methods. (Ref. J. Am. Chem. Soc., 1953, 75, 3843.) Tris(dibenzylideneacetone) dipalladium(0) was purchased from Alfa Aesar. All phosphine ligands were bought from Sigma-Aldrich, Strem, or Alfa Aesar and used as is. All the other reagents and solvents mentioned in this text were bought from Sinopharm Chemical Reagent Co. Ltd or Alfa Aesar and purified when necessary.

^{1}H-NMR, ^{13}C-NMR spectra were recorded on a Bruker Avance 400 spectrometer at ambient temperature in CDCl$_3$ unless otherwise noted. Gas chromatographic (GC) analysis was acquired on a Shimadzu GC-2014 Series GC System equipped with a flame ionization detector. GC-MS analysis was performed on Thermo Scientific AS 3000 Series GC-MS System. HRMS analysis was performed on Finnigan LCQ advantage Max Series MS System. Elementary Analysis was carried out on Elementar Vario EL III elemental analyzer. Organic solutions were concentrated under reduced pressure on a Buchi rotary evaporator. Flash column chromatographic purification of products was accomplished using forced-flow chromatography on Silica Gel (200–300 mesh).

5.4.2 General Procedure

The decarboxylative couplings of potassium pyridyl-2-acetates with diverse aryl bromides, triflates, and activated chlorides.

A 10 ml oven-dried Schlenk tube was charged with $Pd_2(dba)_3$ (0.5–2% mol 3), ligand (1.5–6% mol Xant-Phos), potassium pyridyl-2-acetates (0.36 mmol, 63 mg), and the corresponding aryl electrophile (0.30 mmol) (if solid). The tube was evacuated and filled with argon (this procedure was repeated three times). Then aryl electrophile (0.30 mmol) (if liquid) and diglyme solvent (0.6 mL) were added with a syringe under a counter flow of argon. The tube was sealed with a screw cap, stirred at room temperature for 10 min, and connected to the Schlenk line which was full of argon, stirred in a preheated oil bath (150 °C) for the appointed time (16–24 h). Upon completion of the reaction, the mixture was cooled to room temperature, diluted with ethyl acetate (20 mL) and filtered through a short silica gel column to remove the deposition. The organic layers were washed with water (30 ml × 3), and then with brine, dried over Na_2SO_4, and filtered, the solvents were removed via rotary vapor. Purification of the residue by column chromatography (silica gel, ethyl acetate/hexane gradient) yielded the corresponding products.

5.4.3 Characterization of the Products

Compound name 2-benzylpyridine

49 mg (96%) of a yellow liquid was obtained. This compound is known. Reference: Niwa, T.; Yorimitsu, H.; Oshima, K. Angew. Chem. Int. Ed. 2007, 46, 2643–2645.
^{1}H-NMR (400 MHz, $CDCl_3$, 293K, TMS): δ 4.16 (s, 2H), 7.08–7.11 (m, 2H), 7.19–7.31 (m, 5H), 7.56 (td, J = 1.6, 7.6 Hz, 1H), 8.54 (dd, J = 1.6, 5.6 Hz, 1H).
^{13}C-NMR (100 MHz, $CDCl_3$, 293 K, TMS): δ 44.5, 121.2, 123.1, 126.3, 128.5, 129.0, 136.6, 139.3, 149.1, 160.8.

Compound name 2-(naphthalen-2-ylmethyl)pyridine

57 mg (87%) of a yellow liquid was obtained. This compound is known. Reference: Trost, B. M.; Thaisrivongs, D. A. J. Am. Chem. Soc. 2009, 131, 12056–12057. ^{1}H-NMR (400 MHz, CDCl$_3$, 293K, TMS): δ 4.30 (s, 2H), 7.06–7.11 (m, 2H), 7.36–7.44 (m, 3H), 7.52 (td, J = 1.6, 7.6 Hz, 1H), 7.70–7.78 (m, 4H), 8.55 (d, J = 4.4 Hz, 1H). ^{13}C-NMR (100 MHz, CDCl$_3$, 293 K, TMS): δ 44.8, 121.2, 123.2, 125.4, 125.9, 127.4, 127.50, 127.54, 127.55, 128.1, 132.2, 133.6, 136.5, 136.9, 149.3, 160.8.

Compound name 2-(4-(1,3-dioxolan-2-yl)benzyl)pyridine

63 mg (87%) of a yellow liquid was obtained. This compound is new. ^{1}H-NMR (400 MHz, CDCl$_3$, 293K, TMS): δ 4.00–4.04 (m, 2H), 4.05–4.11 (m, 2H), 4.16 (s, 2H), 5.78 (s, 1H), 7.06–7.10 (m, 2H), 7.27 (d, J = 8.4 Hz, 2H), 7.41 (d, J = 8.0 Hz, 2H), 7.54 (td, J = 2.0, 8.0 Hz, 1H), 8.53 (d, J = 4.4 Hz, 1H). ^{13}C-NMR (100 MHz, CDCl$_3$, 293K, TMS): δ 44.3, 65.2, 103.5, 121.2, 123.0, 126.6, 129.0, 136.0, 136.5, 140.4, 149.2, 160.6. HRMS calcd for C$_{15}$H$_{15}$NO$_2$ (M$^+$) 241.1103; found: 241.1109.

Compound name 2-(thiophen-3-ylmethyl)pyridine

44 mg (83%) of a yellow liquid was obtained. This compound is known. Reference: Trost, B. M.; Thaisrivongs, D. A. J. Am. Chem. Soc. 2009, 131, 12056–12057. ^{1}H-NMR (400 MHz, CDCl$_3$, 293K, TMS): δ 4.16 (s, 2H), 6.96–6.98 (m, 1H), 7.02–7.03 (m, 1H), 7.09–7.13 (m, 2H), 7.24 (dd, J = 2.8, 4.8 Hz, 1H), 7.57 (td, J = 2.0, 8.0 Hz, 1H), 8.53 (d, J = 4.4 Hz, 1H). ^{13}C-NMR (100 MHz, CDCl$_3$, 293K, TMS): δ 39.1, 121.2, 121.7, 122.8, 125.6, 128.4, 136.6, 139.4, 149.2, 160.3.

Compound name 2-(4-(trifluoromethoxy)benzyl)pyridine

70 mg (92%) of a yellow liquid was obtained. This compound is new. ^{1}H-NMR (400 MHz, CDCl$_3$, 293 K, TMS): δ 4.15 (s, 2H), 7.10–7.14 (m, 4H), 7.27 (d, J = 8.4 Hz, 2H), 7.58 (td, J = 2.0, 8.0 Hz, 1H), 8.55 (d, J = 4.4 Hz, 1H). ^{13}C-NMR (100 MHz, CDCl$_3$, 293K, TMS): δ 43.8, 120.5 (q, J_F = 255.1 Hz), 121.0, 121.4, 123.1, 130.3, 136.6, 138.2, 147.8, 149.4, 160.2. HRMS calcd for C$_{13}$H$_{10}$F$_3$NO (M$^+$) 253.0714; found: 253.0708.

Compound name 2-(benzo[d][1,3] dioxol-5-ylmethyl)pyridine

56 mg (88%) of a yellow liquid was obtained. This compound is known. Reference: Bradsher,C.K.; Jones,J.H. J. Org. Chem. 1960, 25, 293–294. ^{1}H-NMR (400 MHz, CDCl$_3$ 293K, TMS): δ 4.06 (s, 2H), 5.89 (s, 2H), 6.73 (m, 3H), 7.08–7.11 (m, 2H), 7.56 (td, J = 2.0, 8.0 Hz, 1H), 8.52–8.54 (dm, J = 4.4 Hz, 1H). ^{13}C-NMR (100 MHz, CDCl$_3$, 293K, TMS): δ 44.2, 100.8, 108.2, 109.4, 121.2, 121.9, 122.9, 133.2, 136.5, 146.0, 147.7, 149.2, 161.0.

Compound name 2-benzyl-4-methylpyridine

34 mg (62%) of a yellow liquid was obtained. This compound is known. Reference: Khatib, S.; Tal, S.; Godsi, O.; Peskin, U.; Eichen, Y. Tetrahedron, 2000, 56, 6753–6762. ^{1}H-NMR (400 MHz, CDCl$_3$ 293 K, TMS): δ 2.27 (s, 3H), 4.13 (s, 2H), 6.93–6.95 (m, 2H), 7.19–7.23 (m, 1H), 7.25–7.31 (m, 4H), 8.38 (d, J = 5.2 Hz, 1H).
    ^{13}C-NMR (100 MHz, CDCl$_3$, 293 K, TMS): δ 21.0, 44.2, 122.4, 124.1, 126.4, 128.6, 129.1, 139.5, 148.1, 148.7, 160.6.

Compound name 2-methyl-6-(4-(trifluoromethyl)benzyl)pyridine

50 mg (67%) of a yellow liquid was obtained. This compound is new. ^{1}H-NMR (400 MHz, CDCl$_3$ 293K, TMS): δ 2.56 (s, 3H), 4.18 (s, 2H), 6.87 (d, J = 7.6 Hz, 1H), 7.00 (d, J = 7.6 Hz, 1H), 7.37 (d, J = 8.0 Hz, 2H), 7.48 (t, J = 7.6 Hz, 1H), 7.54 (d, J = 8.0 Hz, 2H). ^{13}C-NMR (100 MHz, CDCl$_3$, 293 K, TMS): δ 24.3, 44.2, 120.1, 121.2, 124.3 (q, J_F = 270.0 Hz), 125.4 (q, J_F = 3.7 Hz), 128.6 (q, J_F = 32.0 Hz), 129.4, 137.0, 143.6, 158.1, 159.1. HRMS calcd for C$_{14}$H$_{12}$F$_3$N (M$^+$) 251.0922; found: 251.0915.

Compound name 2-benzyl-3-methylpyridine

47 mg (86%) of a yellow liquid was obtained. This compound is known. Reference: Walter et al. J. Heterocycl. Chem. 1977, 14, 47–49. ^{1}H-NMR (400 MHz, CDCl$_3$ 293 K, TMS): δ 2.23 (s, 3H), 4.19 (s, 2H), 7.05–7.08 (m, 1H), 7.14–7.19 (m, 3H), 7.23–7.26 (m, 2H), 7.40 (d, J = 7.6 Hz, 1H), 8.42 (d, J = 3.6 Hz, 1H). ^{13}C-NMR (100 MHz, CDCl$_3$, 293 K, TMS): δ 18.9, 42.2, 121.7, 126.0, 128.3, 128.6, 131.7, 138.0, 139.0, 146.7, 158.7.

Compound name 2-(bis(4-nitrophenyl)methyl)pyridine

38 mg (77%) of a yellow oil was obtained. (No monoarylated product was observed.) ^{1}H-NMR (400 MHz, CDCl$_3$, 293 K, TMS): δ 5.83 (s, 1H), 7.17 (d, J = 8.0 Hz, 1H), 7.24–7.28 (m, 1H), 7.38 (d, J = 8.8 Hz, 4H), 7.71 (td, J = 2.0, 8.0 Hz, 1H), 8.18 (d, J = 8.8 Hz, 4H), 8.63 (d, J = 4.0 Hz, 1H). ^{13}C-NMR (100 MHz, CDCl$_3$, 293K, TMS): δ 58.4, 122.6, 123.9, 124.0, 130.2, 137.3, 147.0, 148.7, 150.1, 160.0.

Compound name 2-(naphthalen-2-ylmethyl)benzo[d]oxazole

44 mg (56%) of a yellow liquid was obtained. This compound is known. Reference: Kumar, R.; Selvam, C.; Kaur, G.; Chakraborti, A. K. Synlett, 2005, 9, 1401–1404.

^{1}H-NMR (400 MHz, CDCl$_3$, 293 K, TMS): δ 4.43 (s, 2H), 7.27–7.30 (m, 2H), 7.44–7.51 (m, 4H), 7.69–7.71 (m, 1H), 7.79–7.83 (m, 4H). ^{13}C-NMR (100 MHz, CDCl$_3$, 293 K, TMS): δ 35.4, 110.5, 119.8, 124.2, 124.8, 125.9, 126.3, 126.9, 127.6, 127.7, 127.8, 128.6, 132.1, 132.6, 133.5, 141.2, 151.0, 165.2.

Compound name 3-(benzo[d]thiazol-2-ylmethyl)benzonitrile

54 mg (72%) of a yellow oil was obtained. This compound is new. ^{1}H-NMR (400 MHz, CDCl$_3$, 293K, TMS): δ 4.46 (s, 2H), 7.37 (t, J = 8.4 Hz, 1H), 7.43–7.50 (m, 2H), 7.56–7.62 (m, 2H), 7.67 (s, 1H), 7.82 (d, J = 8.0 Hz, 1H), 8.01 (d, J = 8.4 Hz, 1H). ^{13}C-NMR (100 MHz, CDCl$_3$, 293K, TMS): δ 39.8, 113.0, 118.5, 121.6, 123.0, 125.2, 126.3, 129.6, 131.1, 132.6, 133.6, 135.5, 138.6, 153.2, 168.6. HRMS calcd for C$_{15}$H$_{10}$N$_2$S (M$^+$): 250.3183; found: 250.3187.

References

1. (a) Baudoin, O. (2007). *Angewandte Chemie International Edition, 46,* 1373–1375. (b) Goossen, L. J., Rodriguez, N., & Goossen, K. (2008). *Angewandte Chemie International Edition, 47,* 3100–3120.
2. (a) Myers, A. G., Tanaka, D., & Mannion, M. R. (2002). *Journal of the American Chemical Society, 124,* 11250–11251. (b) Tanaka, D., & Myers, A. G. (2004). *Organic Letters, 6,*

433–436. (c) Tanaka, D., Romeril, S. P., & Myers, A. G. (2005). *Journal of the American Chemical Society, 127,* 10323–10333.

3. (a) Forgione, P., Brochu, M. C., St-Onge, M., Thesen, K. H., Bailey, M. D., & Bilodeau, F. (2006). *Journal of the American Chemical Society, 128,* 11350–11363. (b) Bilodeau F., Brochu, M. C., Guimond N., Thesen K. H., & Forgione P. (2010). *The Journal of Organic Chemistry, 75,* 1550–1560.

4. (a) Becht, J.-M., & Le Drian, C. (2008). *Organic Letters, 10,* 3161–3164. (b) Becht, J.-M., Catala, C., Le Drian, C., & Wagner, A. (2007). *Organic Letters, 9,* 1781–1783.

5. (a) Maehara, A., Tsurugi, H., Satoh, T., & Miura, M. (2008). *Organic Letters, 10,* 1159–1162. (b) Yamashita, M., Hirano, K., Satoh, T., & Miura, M. (2009). *Organic Letters, 11,* 2337–2340.

6. (a) Moon, J., Jeong, M., Nam, H., Ju, J., Moon, J. H., Jung, H. M., & Lee, S. (2008). *Organic Letters, 10,* 945–948. (b) Moon, J., Jang, M., & Lee, S. (2009). *The Journal of Organic Chemistry, 74,* 1403–1406.

7. (a) Wang, Z. Y., Ding, Q. P., He, X. D., & Wu, J. (2009). *Tetrahedron, 65,* 4635–4638. (b) Voutchkova, A., Coplin, A., Leadbeater, N. E., & Crabtree, R. H. (2008). *Chemical Communications,* 6312–6314. (c) Wang, C., Piel, I., & Glorius, F. (2009). *Journal of the American Chemical Society, 131,* 4194–4195.

8. (a) Shang, R., Fu, Y., Li, J. B., Zhang, S. L., Guo, Q. X., & Liu, L. (2009). *Journal of the American Chemical Society, 131,* 5738–5739. (b) Shang, R., Xu, Q., Jiang, Y.-Y., Wang, Y., & Liu, L. (2010). *Organic Letters, 12,* 1000–1003.

9. (a) Goossen, L. J., Deng, G., & Levy, L. M. (2006). *Science, 313,* 662–664. (b) Goossen, L. J., Rodriguez, N., Melzer, B., Linder, C., Deng, G., & Levy, L. M. (2007). *Journal of the American Chemical Society, 129,* 4824–4833. (c) Goossen, L. J., & Melzer, B. (2007). *The Journal of Organic Chemistry, 72,* 7473–7476. (d) Goossen, L. J., Zimmermann, B., & Knauber, T. (2008). *Angewandte Chemie International Edition, 47,* 7103–7106. (e) Goossen, L. J., & Knauber, T. (2008). *The Journal of Organic Chemistry, 73,* 8631–8634. (f) Goossen, L. J., Rodriguez, N., & Linder, C. (2008). *Journal of the American Chemical Society, 130,* 15248–15249. (g) Goossen, L. J., Manojolinho, F., Khan, B. A., & Rodriguez, N. (2009). *The Journal of Organic Chemistry, 74,* 2620–2623. (h) Goossen, L. J., Rudolphi, F., Oppel, C., & Rodriguez, N. (2008). *Angewandte Chemie International Edition, 47,* 3043–3045. (i) Goossen, L. J., Rodriguez, N., Lange P., & Linder, C. (2010). *Angewandte Chemie International Edition, 49,* 1111–1114.

10. Shang, R., Fu, Y., Wang, Y., Xu, Q., Yu, H.-Z., & Liu, L. (2009). *Angewandte Chemie International Edition, 48,* 9350–9354.

11. (a) Burger, E. C., & Tunge, J. A. (2006). *Journal of the American Chemical Society, 128,* 10002–10003. (b) Waetzig, S. R., & Tunge, J. A. (2007). *Journal of the American Chemical Society, 129,* 4138–4139. (c) Trost, B. M., Xu, J., & Schmidt, T. (2009). *Journal of the American Chemical Society, 131,* 18343–18357. (d) Trost, B. M., Xu, J., & Schmidt, T. (2008). *Journal of the American Chemical Society, 130,* 11852–11853.

12. Decarboxylation of amino acids has been shown to generate electrophilic species: (a) Bi, H.-P., Zhao, L., Liang, Y.-M., & Li, C.-J. (2009). *Angewandte Chemie International Edition, 48,* 792–795. (b) Bi, H.-P., Chen, W.-W., Liang, Y.-M., & Li, C.-J. (2009). *Organic Letters, 11,* 3246–3249. (c) Zhang, C., & Seidel, D. (2010). *Journal of the American Chemical Society, 132,* 1798–1799.

13. (a) Niwa, T., Yorimitsu, H., & Oshima, K. (2007). *Angewandte Chemie International Edition, 46,* 2643–2645. (b) For a related study, see: Qian, B., Guo, S., Shao, J., Zhu, Q., Yang, L., Xia, C., & Huang, H. (2010). *Journal of the American Chemical Society, 132,* 3650–3651.

14. 2- and 4-Pyridylacetic acids decarboxylate thermally in high yield at 90 °C, while the 3-derivative is stable at that temperature; see: Stermitz, F. R., & Huang, W. H. (1971). *Journal of the American Chemical Society, 93,* 3427–3431.

15. Campeau, L. C., Schipper, D. J., & Fagnou, K. (2008). *Journal of the American Chemical Society, 130,* 3266–3267.

16. For related theoretical analysis of decarboxylation and decarboxylative coupling of C(sp^2)-COOH, see: Zhang, S.-L., Fu, Y., Shang, R., Guo, Q.-X., & Liu, L. (2010). *Journal of the American Chemical Society, 132*, 638–646.

17. We isolated the CP2 complex with 4-CN-phenyl substitution according to a previous paper (Yin, J., & Buchwald, S. L. (2002). *Journal of the American Chemical Society, 124*, 6043–6048). We found that CP2 is catalytically active. By addition of CP2 to the reaction between bromobenzene and potassium 2-(2-pyridyl)acetate, we could obtain the desired product (2-benzylpyridine) in 99% GC yield. We also observed a small amount of crossover byproduct, namely, 4-(pyridin-2-ylmethyl)benzonitrile.

18. (a) Kawatsura, M., & Hartwig, J. F. (1999). *Journal of the American Chemical Society, 121*, 1473–1478. (b) Jorgensen, M., Lee, S., Liu, X., Wolkowski, J. P., & Hartwig, J. F. (2002). *Journal of the American Chemical Society, 124*, 12557–12565. (c) Hama, T., Liu, X., Culkin, D. A., & Hartwig, J. F. (2003). *Journal of the American Chemical Society, 125*, 11176–11177. (d) Nguyen, H. N., Huang, X., & Buchwald, S. L. (2003). *Journal of the American Chemical Society, 125*, 11818–11819.

Chapter 6
Synthesis of α-Aryl Nitriles and α-Aryl Acetate Esters Via Palladium-Catalyzed Decarboxylative Couplings of α-Cyano Aliphatic Carboxylate Salts and Malonate Monoester Salts with Aryl Halides

Abstract Palladium-catalyzed decarboxylation coupling reaction between α-cyano aliphatic carboxylate salts with aryl chlorides, aryl bromides, or aryl triflates has been discovered. The reaction is competent for the synthesis of secondary, tertiary, and quaternary α-aryl nitriles utilizing cheap and low-toxic starting materials. This reaction is demonstrated to be useful to synthesize two drug molecules, Flurbiprofen and Anastrozole in gram scale, showcasing its applicability as an alternative choice of the Buchwald–Hartwig α-arylation reaction. The concept of decarboxylative α-arylation reaction was also extended to the synthesis of α-aryl acetate from malonate monoester salt.

6.1 Introduction

α-Aryl nitriles compounds are versatile intermediates in synthetic organic chemistry, because cyano functional group can easily be converted to acid and amide, reduced to primary amine or aldehyde, and form heterocycles via cyclization

© Springer Nature Singapore Pte Ltd. 2017
R. Shang, *New Carbon–Carbon Coupling Reactions Based on Decarboxylation and Iron-Catalyzed C–H Activation*, Springer Theses,
DOI 10.1007/978-981-10-3193-9_6

(Fig. 6.1) [1]. Also, some α-aryl nitriles compounds possess biological activity, such as Anastrozole, which is used to treat breast cancer, as well as Ariflo, Verapamil [2]. In materials science, some molecules that form liquid crystal also contain α-aryl nitriles as structural motifs, as demonstrated by the molecule structure of NCB 807 (Fig. 6.2). Traditional methods to synthesize α-aryl nitrile include the nucleophilic substitution reaction of benzyl halide or benzyl alcohol using potassium cyanide [3], Friedel–Crafts reaction [4], and α-aryl amides dehydration reaction [5]. Recently, Hartwig [6] and Verkade [7] developed new methods to use palladium catalysis to direct α-arylation of nitriles (including 2-cyanoacetate esters) with aryl chlorides and bromides (Fig. 6.3). The requirement of a strong base (e.g., $NaN(SiMe_3)_2$ used to deprotonate the α–C–H bond) in these palladium-catalyzed α-arylation reactions limits the functional group tolerance, and monoarylation is difficult to achieve for acetonitrile and primary nitriles. To solve these two problems, Hartwig and co-workers [8] described improved methods that use relatively expensive α-silyl nitriles and zinc cyanoalkyl reagents to couple with aryl bromides (but not chlorides) (Fig. 6.3). However, for this method, the cost of the coupling is relative high and toxic fluoride is necessary to be used, making the method not suitable for large-scale production. 2-Cyano alkyl zinc reagent is sensitive to air and moisture). We hypothesized that if a palladium (II) intermediate generated after oxidative addition with aryl halide can promote the decarboxylation of α-cyanoacetate smoothly to generate a-cyanomethyl organometallic intermediate in situ, this intermediate may deliver α-aryl nitriles after reductive elimination. Thus, it is possible to achieve a palladium-catalyzed decarboxylative arylation of 2-cyano aliphatic carboxylate salt to synthesize α-aryl nitriles (Fig. 6.4). In this chapter, we report a new type of decarboxylative coupling reaction, namely palladium-catalyzed decarboxylative coupling of aryl bromides, aryl chloride, and aryl triflate with a variety of 2-cyano aliphatic carboxylate salts [9]. 2-Cyano aliphatic carboxylate salts are generally cheap, easily accessible solids that can be easily prepared from corresponding ester via basic hydrolysis. Though Myers [10],

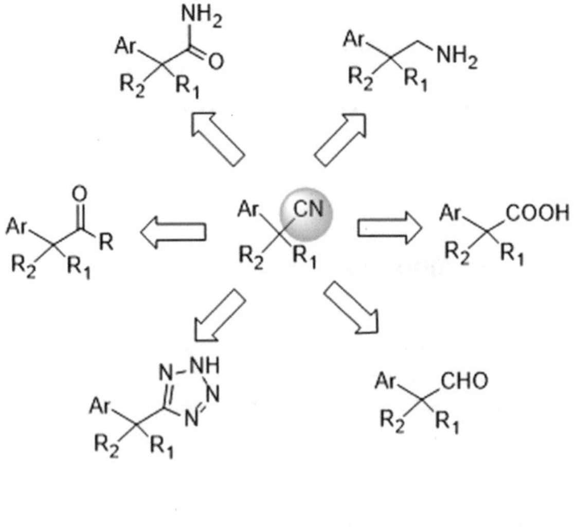

Fig. 6.1 Synthetic chemical conversion of α-aryl nitrile compounds

a-aryl nitrile structures in pharmaceuticals and materials

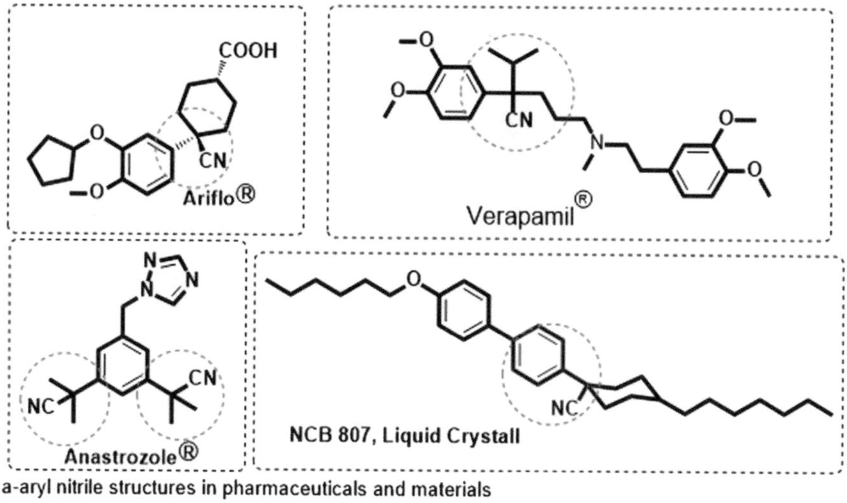

Fig. 6.2 Some drug molecules and molecular materials containing α-aryl nitrile structure

This Work:

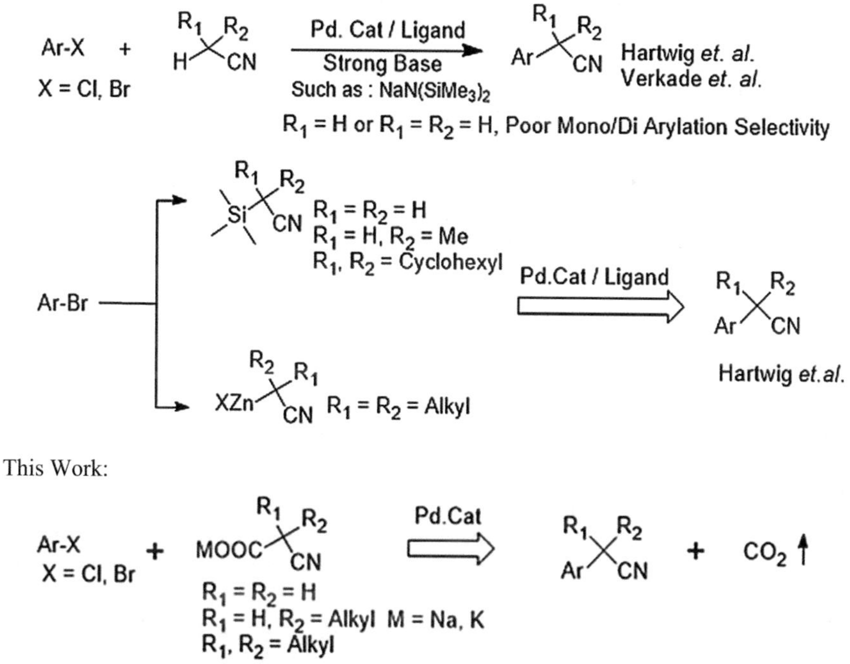

Fig. 6.3 Reported methods for the synthesis of α-aryl nitriles

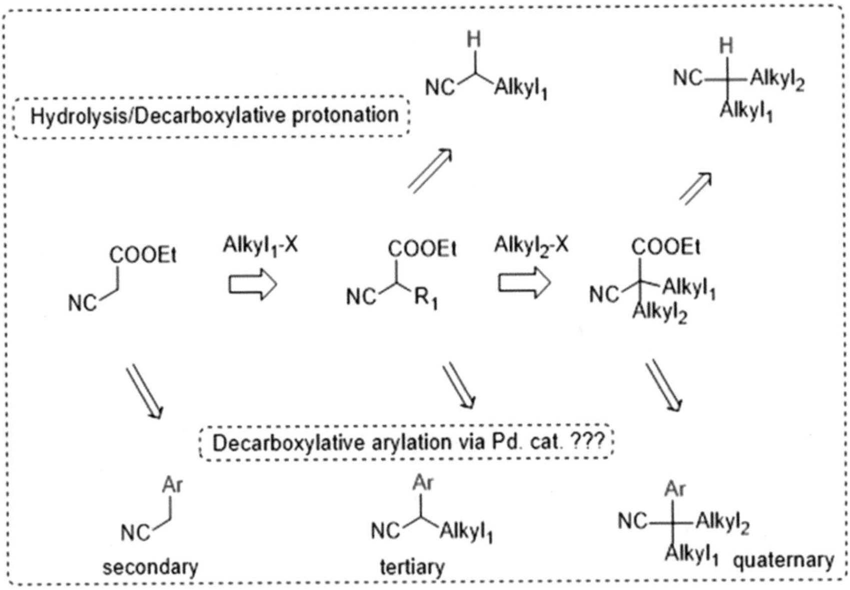

Fig. 6.4 Acid promoted decarboxylative protonation (decarboxylation) and a hypothesis for palladium-catalyzed decarboxylation arylation

Forgione [11], Goossen [12], and the other groups [13–15] have reported a series of decarboxylative coupling reactions, example of decarboxylative arylation to create a tertiary or a quaternary carbon centers is still rare. This work demonstrates that for some synthetic purposes decarboxylative cross-coupling not only provides a conceptually alternative method, but also can be practically favored in terms of both reagent accessibility and reaction scope.

6.2 Results and Discussion

6.2.1 Investigation of the Reaction Conditions

Our study began by testing the coupling of chlorobenzene with sodium or potassium cyanoacetate (Table 6.1). A series of palladium salts, phosphine ligands, and solvents were examined. Under the optimal reaction conditions using [Pd$_2$Cl$_2$(allyl)$_2$]/S-Phos as the catalyst (entry 9), the desired monoarylated product was selectively obtained in 86% yield without the use of any extra base or additives. Interestingly, using potassium cyanoacetate led to the generation of 10% yield of diarylated product **2**, while when sodium cyanoacetate was used, generation of monoarylated product was preferred **1** (entry 8 and 9). The same conditions can also be used for cross-coupling with aryl bromides (entry 15). Under the optimal reaction conditions, without addition of any aryl halide, decarboxylation of

Table 6.1 Decarboxylative coupling under various reaction conditions[a]

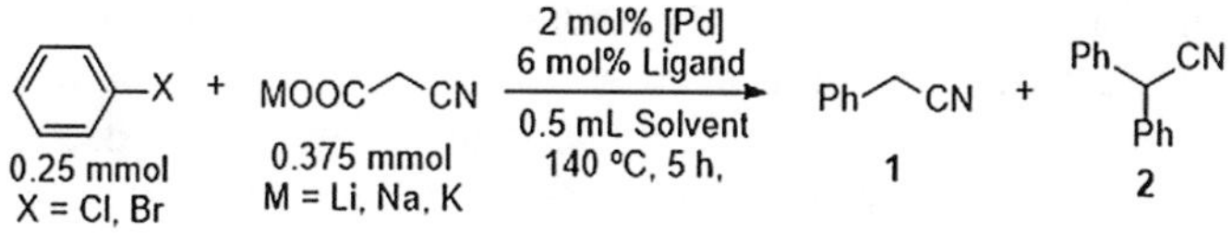

Entry	X, M	[Pd]	Ligand	Solvent	Yield	
					1	2
1	Cl, K	Pd$_2$(dba)$_3$	P(Cy)$_3$	Mesitylene	Trace	16
2	Cl, K	Pd$_2$(dba)$_3$	P(t-Bu)$_3$	Mesitylene	Trace	Trace
3	Cl, K	Pd$_2$(dba)$_3$	X-Phos	Mesitylene	10	28
4	Cl, K	Pd$_2$(dba)$_3$	Cy-JohnPhos	Mesitylene	Trace	21
5	Cl, K	Pd$_2$(dba)$_3$	Dave-Phos	Mesitylene	33	18
6	Cl, K	Pd$_2$(dba)$_3$	Ru-Phos	Mesitylene	65	7
7	Cl, K	Pd$_2$(dba)$_3$	t-BuX-Phos	Mesitylene	18	8
8	Cl, K	Pd$_2$(dba)$_3$	S-Phos	Mesitylene	69	10
9[b]	Cl, Na	[PdCl(allyl)]$_2$	S-Phos	Mesitylene	86	Trace
10	Cl, Li	Pd$_2$(dba)$_3$	S-Phos	Mesitylene	Trace	Trace
11	Cl, Na	Pd$_2$(dba)$_3$	S-Phos	DMA	9	23
12	Cl, Na	Pd$_2$(dba)$_3$	S-Phos	Diglyme	80	6
13[c]	Cl, Na	Pd(OAc)$_2$	S-Phos	Mesitylene	57	13
14[c]	Cl, Na	Pd(TFA)$_2$	S-Phos	Mesitylene	63	4
15	Br, Na	[PdCl(allyl)]$_2$	S-Phos	Mesitylene	88	2

[a]GC yields (average of two runs) using benzophenone as the internal standard. [b]87% isolated yield.
[c]4 mol% [Pd]

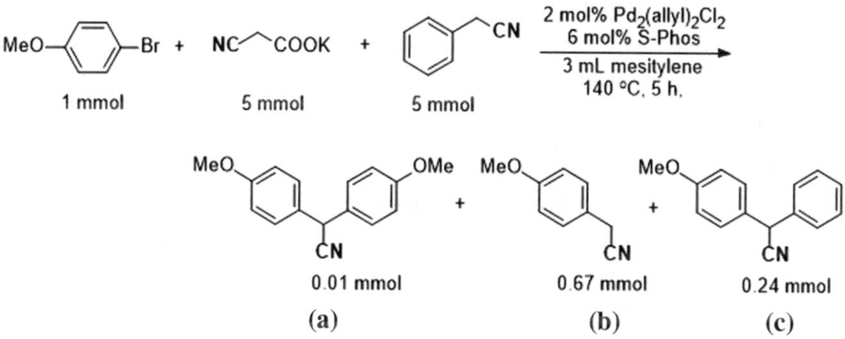

Scheme 6.1 A competition reaction (the amount of product is determined by GC method; which is the average of two runs)

cyanoacetic acid salt to generate acetonitrile was detected (adding acetic acid as a source of protons). In the absence of palladium catalyst, sodium cyanoacetate did not decarboxylate to decompose even at 150 °C. The results of these control experiments indicate the essential role of palladium catalyst for the decarboxylation of cyanoacetate. Moreover, a competition reaction (Scheme 6.1) was conducted, in which 2-phenylacetonitrile was added to the reaction mixture containing 4-bromoanisole and potassium cyanoacetate. We added 2-phenyl acetonitrile to the decarboxylation coupling reaction of potassium cyanoacetate and 4-methoxy bromobenzene, and it was found the main product was product **b**, which was formed via the decarboxylation coupling reaction between potassium cyanoacetate and 4-methoxy bromobenzene. We detected a small amount of diarylated product **a**, we also detected product **c**, formed by the direct α-arylation of phenylacetonitrile with 4-bromoanisole. From the result of this control experiment, it can be concluded the diarylated byproduct shown in Table 6.1 may be formed through the deprotonative arylation of the decarboxylative coupling product.

6.2.2 Exploration of the Substrate Scope

With the optimized reaction condition in hand, we explored the scope of this reaction with respect to aryl halides (Table 6.2). Since the palladium-catalyzed monoarylation of acetonitrile is difficult to achieve due to the uncontrollable arylation, we first focused on decarboxylative arylation of cyanoacetate salt to synthesize α-aryl acetonitriles. The result of the scope studies showed both electron-rich and electron-deficient aryl chlorides react smoothly. A series of electron-rich and electron-deficient aryl bromides can also achieve decarboxylative coupling with sodium cyanoacetate. The functional group compatibility of the reaction is excellent. The reaction is compatible with ether, thioether, fluorine, carbonyl, amide, trimethyl silyl, nitro, cyano, ester, and even the terminal C–C double bond and unprotected amino. Ortho steric hindrance of aryl halides coupling partner is well tolerated, as demonstrated by **1c** and **1l**. Heteroaryl halides (**1t** and **1x**) are also good substrates. Most significantly, base-sensitive functional groups (e.g., ketones and esters with enolizable hydrogen atoms) can be well tolerated in this reaction (**1 k**, **1r**, **1y**). Finally, it is interesting to observe that olefins that can undergo Heck coupling reaction (**1z**) and unprotected amines that are subject to Buchwald amination reaction (**1aa**) can survive the present reaction conditions, showing the orthogonal reactivity of this decarboxylative coupling with traditional cross-coupling reactions.

We further tested the scope of this reaction using tertiary cyanoacetate salts. However, by use of the above protocol, decarboxylative coupling of the 2-cyanopropanoate salt with 4-bromobiphenyl gave the desired product, 2-(biphenyl-4-yl)propanenitrile, in a relatively low yield (47%). Careful analysis of the reaction mixture revealed the formation of biphenyl (40%), thus suggesting the possible occurrence of β-hydride elimination. To solve this problem we changed the

Table 6.2 Decarboxylative coupling of sodium cyanoacetate[a]

$$\text{Ar—X} + \text{NaOOC}\diagup\text{CN} \xrightarrow[\substack{2\text{ mol}\% \text{ Pd}_2(\text{allyl})_2\text{Cl}_2 \\ 6\text{ mol}\% \text{ S-Phos} \\ 140\ ^\circ\text{C, 5 h, 1.0 mL mesitylene}}]{} \text{Ar}\diagup\text{CN} + \text{CO}_2\uparrow$$

0.5 mmol   0.6~0.75 mmol
X = Cl, Br

1a
-Br 84%
-Cl 86%

1b
-Cl 92%

1c
- Cl 92% (*o*-Me)
   97% (*m*-OMe)
   92% (*p*-OMe)

1d
-Br 71%

1e
-Br 91%

1f
-Cl 98%

1g
-Br 95%

1h
-Br 89%

1i
-Br 87%

1j
-Cl 96%
-Br 84%

1l
-Br 88%

1k
-Br 89%

1m
-Cl 60%

1n
-Cl 91%

1o
-Cl 81%

1p
-Cl 88%

1q
-Cl 77%

1r
-Cl 47%

1s
-Cl 87%

1t
-Br 64%

1u
-Cl 86%

1v
-Br 68%

1w
-Br 56%

1x
-Cl 78%

1y
-Br 90%
-Cl 93%

1z
-Cl 85%

1aa
-Cl 78%

1bb
-Cl 48%

[a]Yields based on isolation (%)

ligand to Xantphos, which has been shown to effectively inhibit β-hydride elimination in C–N cross-couplings [16]. To our gratification, the modified protocol can be used to accomplish decarboxylative coupling of tertiary and even quaternary (including cyclic) cyanoacetate salts in good to excellent yields (Tables 6.3 and 6.4). Selective monoarylation is successfully achieved for tertiary cyanoacetate salts, which compares favorably with palladium-catalyzed α-arylation of nitriles under highly basic conditions. Also, base-sensitive groups such as ketones with enolizable hydrogen atoms can survive the reaction conditions. It is important to point out that the tertiary and quaternary cyanoacetate salts can be easily prepared from ethyl cyanoacetate, whereas the corresponding zinc and silicon cyanoalkyl reagents are relatively more expensive to obtain [8]. Furthermore, for decarboxylative arylation of quaternary cyanoacetate salts, it is impossible to proceed via deprotonative arylation followed by decarboxylation. This feature distinguishes the present decarboxylative coupling reaction from the palladium-catalyzed arylation of ethyl cyanoacetate [6, 7].

Table 6.3 Decarboxylative coupling of tertiary cyanoacetates[a]

Reproduced from Angew. Chem. Int. Ed. 2011, 50, 4470 by permission of John Wiley & Sons Ltd.

[a]Isolated yields (%). Note that the potassium salts were used for the reactions, because some sodium salts of tertiary cyanoacetates are hygroscopic and are difficult to be obtained as solid

Table 6.4 Decarboxylative coupling of quaternary cyanoacetates[a]

[a]Isolated yields (%). Note that the potassium salts were used for the reactions, because some sodium salts of quaternary cyanoacetates are hygroscopic and difficult to prepare as solid.
[b]$NaOOCCMe_2CN$ was used

Notably, in the previously reported α-arylation of nitriles [6, 7] as well as arylation of α-silyl nitriles and zinc cyanoalkyl reagents [8], only aryl halides were used as arylation agents. In the present decarboxylative coupling, we were delighted to find that aryl triflates can also be used as the arylation reagents (Table 6.5). This feature additionally expands the utility of the new reaction for the synthesis of α-aryl nitrile, because aryl triflates are more readily accessible from phenols. Interestingly, a slightly lower temperature (120 °C) can be used for the

Table 6.5 Decarboxylative coupling with aryl triflates[a]

Reaction scheme:

$$Ar-OTf + \underset{\substack{\text{0.55-0.75 mmol}}}{\underset{KOOC\quad CN}{\overset{R_1 \quad R_2}{\diagdown}}} \xrightarrow[\substack{120-140\ ^\circ C,\ 10-16\ h \\ 1.0\ mL\ mesitylene}]{\substack{1\text{-}2\ mol\%\ Pd_2(allyl)_2Cl_2 \\ 3\text{-}6\ mol\%\ S\text{-}Phos\ or\ Xant\text{-}Phos}} \underset{Ar\quad CN}{\overset{R_1\quad R_2}{\diagdown}} + CO_2 \uparrow$$

Ar—OTf 0.5 mmol

4a 83%[b]

4b 42%[b]

4c 68%[c]

4d 61%[b]

4e 83%[c]

4f 84%[c]

4g 87%[b]

4h 82%[b]

4i 90%[b]

4j 71%[b]

4k 85%[b]

4l 58%[b]

Reproduced from Angew. Chem. Int. Ed. 2011, 50, 4470 by permission of John Wiley & Sons Ltd.
[a]Isolated yields (%). [b]Xant-Phos was used. [c]S-Phos was used as ligand

decarboxylative coupling with aryl triflates. The reaction can tolerate ortho substitution (**4e** and **4f**) and base-sensitive groups (**4b**). Both electron-rich and electron-poor aryl triflates can be well tolerated, whereas secondary, tertiary, and quaternary cyanoacetate salts are all acceptable coupling partners. In addition, the chloro substitution can be tolerated in the transformation (**4j**), making it possible to accomplish selective sequential cross-coupling reactions.

6.2.3 Application of the Method

In our study, we demonstrated the application of this decarboxylation coupling reaction. We used this decarboxylative coupling reaction to synthesize the Flurbiprofen [17], which is a non-steroidal anti-inflammatory drug (Scheme 6.2). We also used this method to synthesize Anastrozole [2], which is a medicine for the treatment of breast cancer (Scheme 6.3). For the synthesis of Flurbiprofen, the decarboxylation coupling of 2-cyanopropanoate with commercially available 4-bromo-2-fluorobiphenyl affords the nitrile intermediate in 95% yield. Hydrolysis of this nitrile intermediate gives the target compound in 82% yield. For the

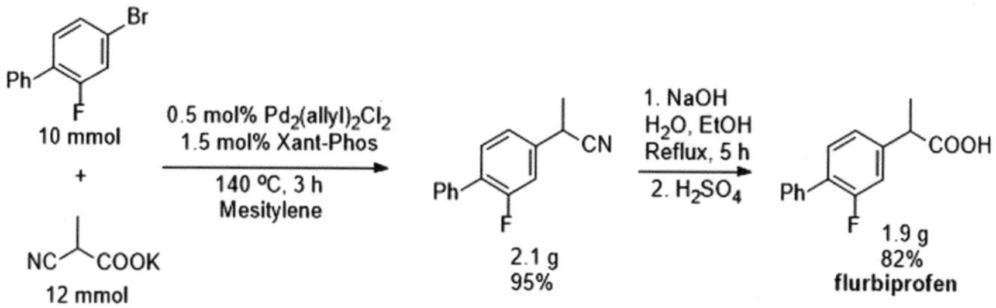

Scheme 6.2 Synthesis of flurbiprofen

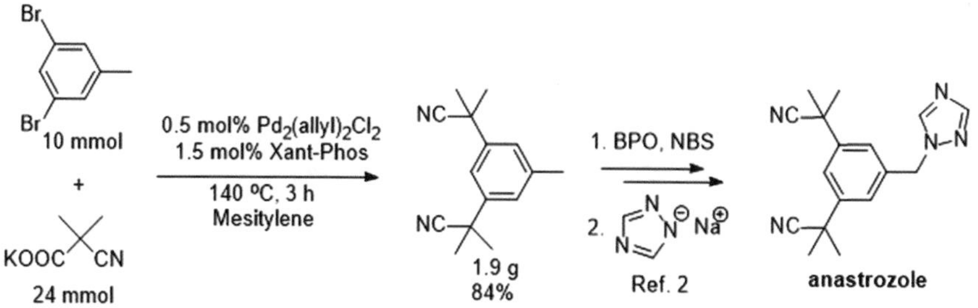

Scheme 6.3 Synthesis of anastrozole. Reproduced from Angew. Chem. Int. Ed. 2011, 50, 4470 by permission of John Wiley & Sons Ltd.

synthesis of Anastrozole, first of all, the decarboxylative coupling of potassium 2-cyano-2-methylpropanoate with commercially available 3,5-dibromotoluene affords the key nitrile intermediate in 84% yield. This intermediate can be readily converted to the target compound through previously established benzylic bromination and substitution reactions [2].

6.2.4 Extending the Concept of Decarboxylative α-Arylation to the Synthesis of α-Aryl Acetates

Finally, it is pointed out that the concept of decarboxylative α-arylation reaction is not limited to nitriles, but also can be extended to carbonyl compounds [18]. For example, in Table 6.6, we demonstrated the palladium-catalyzed decarboxylative coupling of malonate monoester salts with aryl chlorides. This new reaction provides an alternative approach for the synthesis of α-arylated esters that complements the previous palladium-catalyzed α-arylation of esters [19]. This method is also compatible with a series of functional groups, such as ether, thioether, fluorine, amide, nitro, tosyl, cyano, double bond, amines, and heterocyclic groups. No additional base or additive is required in this decarboxylative coupling, while palladium-catalyzed α-aryltion reaction of ester often needs to use strong base (such

Table 6.6 Decarboxylative coupling of malonate monoesters[a]

$$\text{Ar-Cl} \quad + \quad \text{ROOC} \frown \text{COOK} \xrightarrow[\substack{120\text{-}140 \text{ °C, } 10\text{-}20 \text{ h} \\ 1 \text{ mL mesitylene}}]{\substack{2 \text{ mol% } Pd_2(allyl)_2Cl_2 \\ 6 \text{ mol% Ru-Phos}}} \text{Ar} \frown \text{COOR}$$

0.5 mmol   0.75 mmol

6a, 83%
(71%[b])

6b, 71%

6c, 45%

6d, 36%

6e, 57%

6f, 82%

6g, 61%

6h, 38%[c]

6i, 83%[d]

6j, 51%

6k, 65%

6l, 88%

6m, 77%

6n, 60%

6o, 63%

6p, 73%

6q, 85%

6r, 84%

6s, 93%

Reproduced from Angew. Chem. Int. Ed. 2011, 50, 4470 by permission of John Wiley & Sons Ltd.
[a]Yields based on isolation (%). [b]6 mol% S-Phos was used as ligand. [c]0.6 mmol potassium 3-ethoxy-3-oxopropanoate was used. [d]1.5 mmol potassium 3-ethoxy-3-oxopropanoate was used, 36 h

as LiHMDS or LiNCy$_2$) to deprotonate C–H bonds [19]. We believe that the reaction can be expanded to a series of decarboxylative α-arylation reactions, such as the synthesis of α-aryl ketones and α-aryl amides.

6.3   Conclusion

In this chapter, we show the palladium-catalyzed decarboxylative coupling reaction of α-cyanoacetate salt and its derivatives with aryl chlorides, bromides, and even triflates. The reaction use cheap, easily available cyanoacetate salt as raw material for the preparation of diverse α-aryl nitriles with high yield and selectivity. In comparison to the previous methods that use either nitriles or zinc and silicon cyanoalkyl reagents, there are several important advantages of the new reaction including good selectivity towards monoarylation, good functional group tolerance, and good availability of the reactants. Moreover, the concept of α-carbon decarboxylative coupling was extended to carbonyl compounds as an alternative approach for the synthesis of, for instance, α-arylated esters.

6.4 Experiment Section and Compound Data

6.4.1 General Information

All the reactions were carried out in oven-dried Schlenk tubes under argon atmosphere (purity ≥ 99.999%). The mesitylene solvent was bought from Sigma-Aldrich and used without further purification. All aryl halides were bought from Alfa Aesar or Acros and used as is. Aryl triflates and tosylates (*Ref.*:J. Am. Chem. Soc., 2008, 130, 2754) were synthesized according to the literature methods. Potassium cyanoacetate derivatives were synthesized by saponification of the corresponding ethyl ester. The ethyl cyanoacetate derivatives were synthesized by treating ethyl cyanoacetate with corresponding alkyl bromides according to the literature methods (Ref.: Tetrahedron Lett. 1997, 38, 7713–7716). Allyl palladium (II) chloride dimer was purchased from Sigma-Aldrich. All phosphine ligands were bought from Sigma-Aldrich, Strem, Sinocompound, or Alfa Aesar and used as is. All the other reagents and solvents mentioned in this text were bought from Sinopharm Chemical Reagent Co. Ltd or Alfa Aesar and purified when necessary.

^{1}H-NMR, ^{13}C-NMR spectra were recorded on a Bruker Avance 400 spectrometer at ambient temperature in CDCl$_3$ unless otherwise noted. Gas chromatographic (GC) analysis was acquired on a Shimadzu GC-2014 Series GC System equipped with a flame-ionization detector. GC-MS analysis was performed on Thermo Scientific AS 3000 Series GC-MS System. HRMS analysis was performed on Finnigan LCQ advantage Max Series MS System. Elementary Analysis was carried out on Elementar Vario EL III elemental analyzer. Organic solutions were concentrated under reduced pressure on a Buchi rotary evaporator. Flash column chromatographic purification of products was accomplished using forced-flow chromatography on Silica Gel (200–300 mesh).

6.4.2 General Procedures

The decarboxylative couplings of metal cyanoacetates with phenylhalides.

A 10 mL oven-dried Schlenk tube was charged with Pd source (see Table 6.1), ligand (see Table 6.1), and metal cyanoacetate (see Table 6.1, 0.375 mmol). The tube was evacuated and filled with argon (this procedure was repeated three times).

Then phenyl halide (0.25 mmol) and solvent (0.5 mL, see Table 6.1) were added with a syringe under a counter flow of argon. The tube was sealed with a screw cap, stirred at room temperature for 10 min, and connected to the Schlenk line that was full of argon, stirred in a preheated oil bath (140 °C) for the appointed time (5 h). Upon completion of the reaction, the mixture was cooled to room temperature and diluted with diethyl ether, and the yields were determined by gas chromatography using benzophenone as the internal standard.

6.4.3 Characterization of the Products

Compound name 2-phenylacetonitrile

51 mg (87%) of a colorless liquid was obtained. This compound is known. Reference: Kosugi, M.; Ishiguro, M.; Negishi, Y.; Sano, H.; Migita, T. *Chem. Lett.* 1984, *13*, 1511–1512. ^{1}H-NMR (400 MHz, CDCl$_3$, 293 K, TMS): δ 3.71 (s, 2H), 7.30–7.38 (m, 5H). ^{13}C-NMR (100 MHz, CDCl$_3$, 293 K, TMS): δ 23.4, 117.8, 127.8, 127.9, 129.0, 129.8. HRMS calcd for C$_8$H$_7$N (M+) 117.0578; found: 117.0575.

Compound name 2-(4-(trifluoromethyl)phenyl)acetonitrile

56 mg (60%) of a colorless liquid was obtained. This compound is known. Reference: Wu, L.; Hartwig, J. F. *J. Am. Chem. Soc.* 2005, *127*, 15824–15832. ^{1}H-NMR (400 MHz, CDCl$_3$, 293 K, TMS): δ 3.83 (s, 2H), 7.47 (d, J = 8.0 Hz, 2H), 7.65 (d, J = 8.0 Hz, 2H). ^{13}C-NMR (100 MHz, CDCl$_3$, 293 K, TMS): δ 23.4, 116.9, 123.7 (q, J_F = 270.5 Hz), 126.1 (q, J_F = 3.8 Hz), 128.3, 130.5 (q, J_F = 33.0 Hz), 133.9. HRMS calcd for C$_9$H$_6$F$_3$N (M+) 185.0452; found: 185.0455.

Compound name isopropyl 3-(cyanomethyl)benzoate

92 mg (91%) of a colorless oil was obtained. This compound is new. ^{1}H-NMR (400 MHz, CDCl$_3$, 293 K, TMS): δ 1.38 (d, J = 6.0 Hz, 6H), 3.81 (s, 2H), 5.22–5.31 (m, 1H), 7.47 (t, J = 7.6 Hz, 1H), 7.54 (d, J = 8.0 Hz, 1H), 7.98 (s, 1H), 8.00 (d, J = 7.6 Hz, 1H). ^{13}C-NMR (100 MHz, CDCl$_3$, 293 K, TMS): 21.8, 23.4, 68.8, 117.3, 129.0, 129.15, 129.19, 130.2, 131.8, 132.0, 165.3. HRMS calcd for C$_{12}$H$_{13}$NO$_2$ (M+) 203.0946; found: 203.0947.

Compound name 2-(3-(morpholine-4-carbonyl)phenyl)acetonitrile

100 mg (87%) of a colorless oil was obtained. This compound is new. ^{1}H-NMR (400 MHz, CDCl$_3$, 293 K, TMS): δ 3.44–3.78 (br, 8H), 3.80 (s, 2H), 7.35–7.46 (m, 4H). ^{13}C-NMR (100 MHz, CDCl$_3$, 293 K, TMS): δ 23.2, 42.4, 48.0, 66.6, 117.3, 126.4, 126.6, 129.2, 130.6, 136.0, 169.3. HRMS calcd for C$_{13}$H$_{14}$N$_2$O$_2$ (M+) 230.1055; found: 230.1058.

Compound name 2-(thiophen-3-yl)acetonitrile

48 mg (78%) of a yellow liquid was obtained. This compound is known. Reference: Campaigne; McCarthy; *J. Am. Chem. Soc.* 1954, *76*, 4466. ^{1}H-NMR (400 MHz, CDCl$_3$, 293 K, TMS): δ 3.73 (s, 2H), 7.02 (dd, J_1 = 5.2 Hz, J_2 = 1.2 Hz, 1H), 7.25–7.26 (m, 1H), 7.36 (dd, J_1 = 5.2 Hz, J_2 = 3.2 Hz, 1H). ^{13}C-NMR (100 MHz, CDCl$_3$, 293 K, TMS): δ 18.8, 117.6, 123.1, 127.0, 127.2, 129.4. HRMS calcd for C$_6$H$_5$NS (M+) 123.0143; found: 123.0143.

Compound name 4-(1-cyanoethyl)phenyl 4-methylbenzenesulfonate

134 mg (89%) of a yellow solid was obtained. This compound is new. ^{1}H-NMR (400 MHz, CDCl$_3$, 293 K, TMS): δ1.61 (d, J = 7.2 Hz, 3H), 2.46 (s, 3H), 3.89 (q, J = 7.2 Hz, 1H), 7.00 (d, J = 8.4 Hz, 2H), 7.28 (d, J = 8.8 Hz, 2H), 7.33 (d, J = 8.0 Hz, 2H), 7.70 (d, J = 8.0 Hz, 2H). ^{13}C-NMR (100 MHz, CDCl$_3$, 293 K, TMS): δ 21.2, 21.6, 30.6, 121.0, 123.0, 128.0, 128.4, 129.8, 132.1, 135.9, 145.6, 149.1. HRMS calcd for C$_{16}$H$_{15}$NO$_3$S (M+) 301.0773; found: 301.0771.

Compound name ethyl 4-(2-cyanobutan-2-yl)benzoate

73 mg (63%) of a yellow oil was obtained. This compound is new. ^{1}H-NMR (400 MHz, CDCl$_3$, 293 K, TMS): δ 0.96 (t, J = 7.4 Hz, 3H), 1.40 (t, J = 7.2 Hz, 3H), 1.73 (s, 3H), 1.99 (m, 2H), 4.39 (q, J = 7.2 Hz, 2H), 7.51 (d, J = 8.8 Hz, 2H), 8.06 (d, J = 8.8 Hz, 2H). ^{13}C-NMR (100 MHz, CDCl$_3$, 293 K, TMS): δ 9.8, 14.3, 27.2, 35.2, 43.4, 61.1, 122.8, 125.7, 130.12, 130.15, 145.0, 166.0. HRMS calcd for C$_{14}$H$_{17}$NO$_2$ (M+) 231.1259; found: 231.1260.

Compound name 1-(4-trifluoromethyl-phenyl)-cyclohexanecarbonitrile

108 mg (85%) of a yellow liquid was obtained. ^{1}H-NMR (400 MHz, CDCl$_3$, 293 K, TMS) δ 1.22–1.34 (m, 1H), 1.74–1.93 (m, 7H), 2.16 (d, J = 11.4 Hz, 2H), 7.60–7.68 (m, 4H). ^{13}C-NMR (100 MHz, CDCl$_3$, 293 K, TMS) δ 23.4, 24.8, 37.2, 44.5, 121.9, 123.8 (q, J_F = 272.1 Hz), 125.9 (q, J_F = 3.7 Hz), 126.1, 130.2 (q, J_F = 32.6 Hz), 145.3. HRMS calcd for C$_{14}$H$_{14}$F$_3$N (M+) 253.1078; found: 253.1072.

Compound name 1-(4-(trimethylsilyl)phenyl)cyclohexanecarbonitrile

117 mg (91%) of a yellow solid was obtained. ^{1}H-NMR (400 MHz, CDCl$_3$, 293 K, TMS) δ 0.25–0.28 (m, 9H), 1.23–1.34 (m, 1H), 1.73–1.88 (m, 7H), 2.14 (d, J = 11.6, 2H), 7.45–7.49 (m, 2H), 7.52–7.56 (m, 2H). ^{13}C-NMR (100 MHz, CDCl$_3$, 293 K, TMS) δ -1.2, 23.6, 25.0, 37.2, 44.3, 122.6, 124.9, 133.9, 140.2, 141.9. HRMS calcd for C$_{16}$H$_{23}$NSi (M+) 257.1600; found: 257.2597.

Compound name (4-trimethylsilanyl-phenyl)acetonitrile

67 mg (71%) of a yellow liquid was obtained. ^{1}H-NMR (400 MHz, CDCl$_3$, 293 K, TMS) δ 0.27 (s, 9H), 3.73 (s, 2H), 7.31 (d, J = 8.0 Hz, 2H), 7.53 (d, J = 8.0 Hz, 2H). ^{13}C-NMR (100 MHz, CDCl$_3$, 293 K, TMS) δ -1.3, 23.5, 117.7, 127.2, 130.3, 134.1, 140.6. HRMS calcd for C$_{11}$H$_{15}$NSi (M+) 189.0974; found: 189.0968.

Compound name (4-cyanomethyl-phenyl)-acetic acid ethyl ester

90 mg (89%) of a yellow liquid was obtained. ^{1}H-NMR (400 MHz, CDCl$_3$, 293 K, TMS) δ 1.25 (t, J = 7.1 Hz, 3H), 3.61 (s, 2H), 3.71 (s, 2H), 4.14 (q, J = 7.1 Hz, 2H), 7.26–7.31 (m, 4H). ^{13}C-NMR (100 MHz, CDCl$_3$, 293 K, TMS) δ 14.0, 23.1, 40.8 60.8, 117.7, 128.0, 128.6, 129.9, 134.0, 171.1. HRMS calcd for C$_{12}$H$_{13}$NO$_2$ (M+) 203.0946; found: 203.0941.

Compound name (3-imidazol-1-yl-phenyl)-acetic acid ethyl ester

89 mg (77%) of a yellow liquid was obtained. **¹H-NMR** (400 MHz, CDCl₃) δ 1.27
(t, *J* = 7.1 Hz, 3H), 3.68 (s, 2H), 4.18 (q, *J* = 7.2 Hz, 2H), 7.20 (s, 1H), 7.27–7.36
(m, 4H), 7.43 (t, *J* = 7.8 Hz, 1H), 7.86 (s, 1H). ¹³C-NMR (100 MHz, CDCl₃) δ
14.1, 40.9, 61.0, 118.1, 120.1, 122.3, 128.3, 129.9, 130.3, 135.5, 136.1, 137.4,
170.8. HRMS calcd for $C_{13}H_{14}N_2O_2$ (M⁺) 230.1055; found: 230.1052.

References

1. Friedrich, K., & Wallenfels, K. (1970). *The Chemistry of the Cyano Group;* *Wiley-Interscience: New York.*
2. Alnabari, M., Freger, B., Arad, O., Zelikovitch, L., Seryi, Y., Danon, E., Davidi, G., & Kaspi, J. (2006). *U.S. Patent 2006/0035950 A1.*
3. Soli, E. D., Manoso, A. S., Patterson, M. C., Deshong, P., Favor, D. A., Hirschmann, R., & Smith, A. B. (1999). *The Journal of Organic Chemistry, 64,* 3171–3177.
4. Kurz, M. E., Lapin, S. C., Mariam, A., Hagen, T. J., & Qian, X. Q. (1984). *The Journal of Organic Chemistry, 49,* 2728–2733.
5. Narsaiah, A. V., & Nagaiah, K. (2004). *Advanced Synthesis & Catalysis, 346,* 1271–1274.
6. Stauffer, S. R, Beare, N. A., Stambuli, J. P., & Hartwig, J. F. (2001). *Journal of the American Chemical Society, 123,* 4641–4642.
7. You, J., & Verkade, J. G. (2003). *Angewandte Chemie International Edition, 42,* 5051–5053.
8. Wu, L., & Hartwig, J. F. (2005). *Journal of the American Chemical Society, 127,* 15824–15832.
9. Recio III, A., & Tunge, J. A. (2009). *Organic Letters, 11,* 5630–5633.
10. (a) Myers, A. G., Tanaka, D., & Mannion, M. R. (2002). *Journal of the American Chemical Society, 124,* 11250–11251.
11. Forgione, P., Brochu, M. C., St-Onge, M., Thesen, K. H., Bailey, M. D., & Bilodeau, F. (2006). *Journal of the American Chemical Society, 128,* 11350–11351.
12. (a) Goossen, L. J., Deng, G., & Levy, L. M. (2006). *Science, 313,* 662–664. (b) Goossen, L. J., Rodriguez, N., Melzer, B., Linder, C., Deng, G., & Levy, L. M. (2007). *Journal of the American Chemical Society, 129,* 4824–4833. (c) Goossen, L. J., Rodriguez, N., & Linder, C. (2008). *Journal of the American Chemical Society, 130,* 15248–15249.
13. (a) Becht, J.-M., Catala, C., Le Drian, C., & Wagner, A. (2007). *Organic Letters, 9,* 1781–1783. (b) Wang, C., Piel, I., & Glorius, F. (2009). *Journal of the American Chemical Society, 131,* 4194–4195.
14. (a) Burger, E. C., & Tunge, J. A. (2006). *Journal of the American Chemical Society, 128,* 10002–10003. (b) Waetzig, S. R., & Tunge, J. A. (2007). *Journal of the American Chemical Society, 129,* 4138–4139.
15. (a) Shang, R., Fu, Y., Li, J. B., Zhang, S. L., Guo, Q. X., & Liu, L. (2009). *Journal of the American Chemical Society, 131,* 5738–5739. (b) Shang, R., Fu, Y., Wang, Y., Xu, Q., Yu, H.-Z., & Liu, L. (2009). *Angewandte Chemie International Edition, 48,* 9350–9454. (c) Zhang, S.-L., Fu, Y., Shang, R., Guo, Q.-X., & Liu, L. (2010). *Journal of the American Chemical Society, 132,* 638–646. (d) Shang, R., Yang, Z. W., Wang, Y., Zhang, S. L., & Liu, L. (2010). *Journal of the American Chemical Society, 132,* 14391–14393.

16. Peng, J., Lin, W., Yuan, S., & Chen, Y. (2007). *The Journal of Organic Chemistry, 72*, 3145–3148.
17. Quasdorf, K. W., Riener, M., Petrova, K. V., & Garg, N. K. (2009). *Journal of the American Chemical Society, 131*, 17748–17749.
18. (a) Culkin, D. A., & Hartwig, J. F. (2003). *Accounts Chemical Reseach, 36*, 234–245. (b) (c) Hama, T., Liu, X., Culkin, D. A., & Hartwig, J. F. (2003). *Journal of the American Chemical Society, 125*, 11176–11177. (c) Bellina, F., & Rossi, R. (2010). *Chemical Reviews, 110*, 1082–1146.
19. Jorgensen, M., Lee, S., Liu, X., Wolkowski, J. P., & Hartwig, J. F. (2002). *Journal of the American Chemical Society, 124*, 12557–12565.

Chapter 7
Palladium-Catalyzed Decarboxylative Couplings of Nitrophenyl Acetate Salts and Its Derivatives with Aryl Halides

Abstract In this chapter, we report the palladium-catalyzed decarboxylative cross-coupling of 2-nitrophenyl acetates and 4-nitrophenyl acetates with aryl bromides and chlorides. Because the nitro group can be easily converted to many other functional groups in synthetic organic chemistry, this reaction provides a new path for the synthesis of diverse functionalized 1,1-diaryl methanes and their derivatives.

7.1 Introduction

Transition-metal-catalyzed decarboxylative cross-coupling has emerged as a useful method for C–C bond formation recently [1]. Using carboxylic acids as alternative reagents over organometallic compounds, this type of transformation avoids strong basic reaction conditions and generates nontoxic CO_2 as the byproduct. Pioneering studies by Myers [2], Forgione [3], and Goossen [4] focused on the intermolecular decarboxylative coupling reactions of aromatic, alkenyl, and alkynyl acids [5]. Tunge [6], Trost [7], Stoltz [8], and others [9] studied palladium-catalyzed

© Springer Nature Singapore Pte Ltd. 2017
R. Shang, *New Carbon–Carbon Coupling Reactions Based on Decarboxylation and Iron-Catalyzed C–H Activation*, Springer Theses,
DOI 10.1007/978-981-10-3193-9_7

intramolecular decarboxylative allylation and benzylation reactions. Although this type of the reaction has been widely reported in the last few years, most studies focused on $C(sp^2)$-COOH, relatively less example has been reported on the intermolecular decarboxylative coupling of aliphatic carboxylic acids [10].

In Chaps. 5 and 6, we described Pd-catalyzed decarboxylative couplings of the alkali salts of 2-(2-azaaryl) acetates [11] and cyanoacetates [12] with aryl halides. These reactions not only provide conceptually alternative methods for the preparation of functionalized azaarenes and 2-arylnitriles, but also demonstrate the practical value in decarboxylative coupling reactions in terms of both reagents availability and reaction scopes. In this chapter, we report a new example of this family of reactions, namely, the decarboxylative cross-coupling of potassium 2- and 4-nitrophenyl acetates with aryl halides. Due to the versatile conversion of the nitro group to many other functional groups in the synthetic chemistry [13], this reaction provides a useful method for the preparation of diverse 1,1-diaryl methanes and their derivatives [14, 15]. Note that Waetzig and Tunge previously described the Pd-catalyzed decarboxylative allylation of nitrophenyl acetates [6e].

7.2   Results and Discussion

7.2.1   Investigation of the Reaction Conditions

Our work started with testing decarboxylative coupling of potassium 2-nitrophenyl acetate with chlorobenzene as model reaction to search the optimal reaction condition (Table 7.1). Using [PdCl(allyl)]$_2$ and P(Ad)$_2^n$Bu as the ligand, the desired transformation took place readily to give the coupling product **1** in good yield (entry 1), while simultaneously decarboxylative protonation of 2-nitrophenyl acetate also occurred to produce byproduct 2. Interestingly, when P(^{t}Bu)$_3$ or ^{t}BuX-Phos was chosen as the ligand, this reaction does not proceed (entries 2–3), which may be caused by the large steric hindrance exerted by the tert-butyl group on the ligand. S-Phos, Dave-Phos, DCyPF, and Brett-Phos promoted the coupling to various extents (entries 4–7). The optimal ligand turned out to be X-Phos, and the corresponding yield was 91% (entry 8). Note that X-Phos can also be used with Pd$_2$(dba)$_3$ as the catalyst in the decarboxylative coupling reaction and the yield was 87% (entry 9), but when Pd(OAc)$_2$ was used, the decarboxylative coupling could not proceed smoothly generating mainly product of decarboxylative protonation (entry 10). Testing potassium 4-nitrophenyl acetate under the optimized reaction condition revealed potassium 4-nitrophenyl acetate was also amenable (entry 11). However, if potassium 3-nitrophenyl acetate was used as the substrate, neither decarboxylative coupling nor decarboxylative protonation took place, indicating that decarboxylation of potassium 3-nitrophenylacetate cannot occur under the optimal condition (entry 12). Furthermore, bromobenzene can be used in the decarboxylative coupling with the potassium salts of both 2- and 4-nitrophenyl

Table 7.1 Pd-catalyzed decarboxylative coupling of potassium nitrophenyl acetates with phenyl halides[a]

Entry	X	Position	[Pd]	Ligand	GC yield (%)	
					1	2
1	Cl	2-NO$_2$	[PdCl(allyl)]$_2$	P(Ad)$_2^n$Bu	83	18
2	Cl	2-NO$_2$	[PdCl(allyl)]$_2$	P(^{t}Bu)$_3$	<5	61
3	Cl	2-NO$_2$	[PdCl(allyl)]$_2$	t-BuX-Phos	<5	74
4	Cl	2-NO$_2$	[PdCl(allyl)]$_2$	S-Phos	85	16
5	Cl	2-NO$_2$	[PdCl(allyl)]$_2$	Dave-Phos	83	17
6	Cl	2-NO$_2$	[PdCl(allyl)]$_2$	DCyPF	51	43
7	Cl	2-NO$_2$	[PdCl(allyl)]$_2$	Brett-Phos	56	36
8	Cl	2-NO$_2$	[PdCl(allyl)]$_2$	X-Phos	91	16
9	Cl	2-NO$_2$	Pd$_2$(dba)$_3$	X-Phos	87	21
10[b]	Cl	2-NO$_2$	Pd(OAc)$_2$	X-Phos	7	73
11	Cl	4-NO$_2$	[PdCl(allyl)]$_2$	X-Phos	89	16
12	Cl	3-NO$_2$	[PdCl(allyl)]$_2$	X-Phos	n.r.	<5
13	Br	2-NO$_2$	[PdCl(allyl)]$_2$	X-Phos	85	18
14	Br	4-NO$_2$	[PdCl(allyl)]$_2$	X-Phos	87	19

[a]All the reactions were carried out in 0.25 mmol scale in 0.5 mL mesitylene. GC yields are measured as average of two runs, using benzophenone as internal standard. [b]4% mol [Pd]

acetates (entries 13–14). Note that addition of radical scavenger, such as 1,1-diphenylethylene, to the reaction system did not inhibit the reaction, suggesting the reaction does not proceed via a radical mechanism.

7.2.2 Exploration of Substrate Scope

To further explore the substrate scope, we examined decarboxylative coupling between a variety of aryl halides and nitrophenyl acetate (Table 7.2). Both electron-rich and electron-poor aryl chlorides as well as bromides can be successfully converted across a wide range of functional groups including ether (**3e, 4g, 4i**), ester (**3c, 3r**), tosyl (**3a**), fluoro (**3d**), trifluoromethyl (**3f**), thioether (**3p**), silyl (**3l, 4h**), nitrile (**3b, 3h, 3n**), olefin (**3k**), acetal (**3q**), and amine (**3g, 3i**). The isolated yields ranged from modest to excellent (>80%). It is notable that, for some electron-poor aryl halides, S-Phos has to be used to replace X-Phos to improve the reaction yields. Substitution at the *ortho*-position (**4d**) can be tolerated. Halogenated

Table 7.2 Aryl halide scope[a]

$$\underset{\substack{\text{0.6 mmol}}}{O_2N\text{-}\text{C}_6\text{H}_4\text{-}CH_2COOK} + \underset{\substack{\text{0.5 mmol}}}{(Het)Ar\text{-}X} \xrightarrow[\substack{1\text{ mL mesitylene}\\140\ ^\circ C,\ 5\text{-}10\ h}]{\substack{1\text{-}2\text{ mol\% } Pd_2(allyl)_2Cl_2\\3\text{-}6\text{ mol\% X- Phos}}} O_2N\text{-}C_6H_4\text{-}CH_2\text{-}(Het)Ar$$

X = Cl

3a, R = OTs[b], 80%
3b, R = CN[b], 63%
3c, R = CO$_2$Et[b], 68%
3d, R = F[b], 94%
3e, R = OMe, 95%
3f, R = CF$_3$[b], 76%

X = Br
3g, R = NMe$_2$, 86%

3h, X = Cl, 72%[b]

3i, X = Cl, 88%

3j, X = Br, 91%

3k, X = Cl, 82%

3l, X = Br, 89%

3m, X = Cl, 92%

3n, X = Cl 86%[b]

3o, X = Cl, 42%

3p, X = Cl, 94%

3q, X = Br, 95%

3r, X = Br, 86%

4a, X = Cl, 94%

4b, X = Cl, 96%

4c, X = Br, 63%

4d, X = Cl, 55%

4e, X = Br, 58%

4f, X = Br, 55%

4g, X = Cl, 83%

4h, X = Br, 85%

4i, X = Br, 75%

4j, X = Cl, 64%

4k, X = Cl, 55%

Reprinted with the permission from Org. Lett. 2011, 13, 4240. Copyright 2011 American Chemical Society

[a]All the reactions were carried out at 0.50 mmol scale in 1.0 mL mesitylene. The yield (average of two runs) of the isolated product was calculated based on the quantity of the aryl halide. [b]S-Phos was used as ligand

heteroarenes, especially those containing nitrogen atoms such as benzothiazole (**3m**), thiophene (**3o**), indole (**4c**), pyridine (**4k**), and quinoline (**4f, 4j**) are also good substrates. Finally, for those substrates that possess both the aryl–Cl bond and the aryl–Br bond, the aryl–Br bond can be selectively cross-coupled in the presence of the aryl–Cl bond (**4e**).

In the above reactions, the carboxylate reagents are easily generated from commercially available 2- and 4-nitrophenyl acetic acids. According to the previous studies [16], the potassium salts of α-alkylated 2- and 4-nitrophenyl acetates can also be readily prepared through the simple alkylation reactions of 2- and 4-nitrophenyl acetate esters. Gratifyingly, we found that these substituted potassium carboxylates can also react with various aryl halides through decarboxylative cross-coupling (Table 7.3). In Table 7.3, we can see that both α-alkylated and α-benzyl nitrophenyl acetates could be the suitable substrates. These reactions again tolerate diverse functional groups including ether, ester, fluoro, trifluoromethyl, nitrile, acetal, and amine and heterocycles such as pyridine and benzothiazole. It is interesting to notice that α-dialkylated substrates (**5i, 5j, 5k**) can also undergo the decarboxylative coupling to generate quaternary carbon centers. The reactions of these particular substrates suggest that decarboxylation takes place prior to cross-coupling.

7.2.3 Application of the Reaction in Synthesis of 4-Aryl Quinolines and Dihydroquinolinones

The above decarboxylative coupling reactions can be used to synthesize a variety of nitro-substituted 1,1-diarylmethanes. The synthetic utility of this method may be substantially increased when the NO_2 substituent is converted to other groups [13]. Scheme 7.1 shows a new synthesis of 4-aryl quinolines and dihydroquinolinones. By reducing NO_2 to NH_2 using either $SnCl_2$ or Zn/HCl, followed by condensation, 4-aryl quinolines and dihydroquinolinones could be readily synthesized. In the previous catalytic allylation of nitropheny lacetates, Tunge et al. also showed that decarboxylative coupling could be followed by reductive cyclization to afford dihydroquinolones or quinolones [6e].

7.2.4 Transformation of -NO₂ by Sandmeyer Reaction

Importantly, NH_2, which is reduced from NO_2, can be further converted to many other groups through the Sandmeyer reaction. Therefore, the above decarboxylative coupling reactions can be used to prepare a variety of 1,1-diaryl methanes. Scheme 7.2 shows that a 2-benzylated phenol can be synthesized by using the

Table 7.3 Coupling of R-Alkylated Substrates[a]

5a, X = Br, 73%[b]   **5b**, X = Br, 63%[b]   **5c**, X = Br, 55%[b]

5d, X = Cl, 95%   **5e**, X = Br, 95%   **5f**, X = Cl, 73%
X = Br, 95%[b]

5g, X = Br, 95%   **5h**, X = Cl, 89%

5i, X = Cl, 32%   **5j**, X = Br, 48 %   **5k**, X = Cl, 28%

6a, X = Cl, 95%   **6b**, X = Cl, 70%[c]   **6c**, X = Cl, 85%
X = Br, 96%[b]

6d, X = Cl, 94%   **6e**, X = Cl, 92%   **6f**, X = Br, 87%

6g, X = Cl, 80%   **6h**, X = Cl, 91%

Reprinted with the permission from Org. Lett. 2011, 13, 4240. Copyright 2011 American Chemical Society

[a]All the reactions were carried out at 0.50 mmol scale in 1.0 mL mesitylene. The yield (average of two runs) of the isolated product was calculated based on the quantity of the aryl halide. [b]Xant-Phos was used as ligand. [c]S-Phos was used as ligand

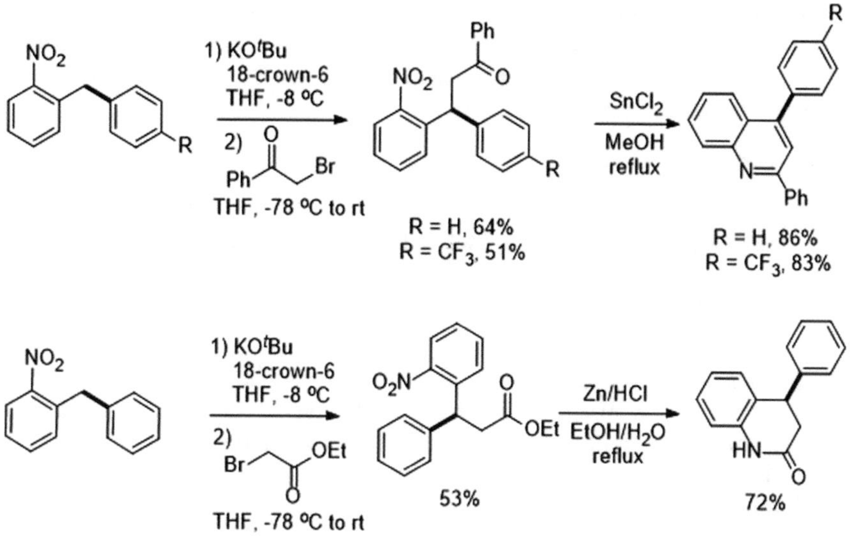

Scheme 7.1 New synthesis of 4-Aryl quinolines and dihydroquinolinones. Reprinted with the permission from Org. Lett. **2011**, 13, 4240. Copyright 2011 American Chemical Society

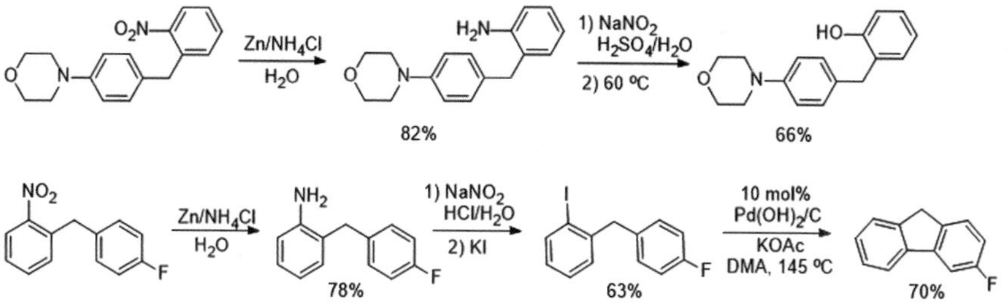

Scheme 7.2 Further transformations by using Sandmeyer reaction

decarboxylative cross-coupling. Furthermore, by reduction of -NO_2 to -NH_2, and followed by the Sandmeyer iodination reaction, functionalized fluorene could be synthesized from the iodide in the presence of a palladium catalyst [17].

7.3 Conclusion

In summary, we reported a practical protocol for the palladium-catalyzed decarboxylative cross-coupling of potassium 2- and 4-nitrophenyl acetates with aryl chlorides and bromides. This reaction adds a new example to the repertoire of catalytic decarboxylative cross-coupling of activated aliphatic carboxylic acids [18]. Because the nitro group can be readily converted to various other functional groups, the new reaction provides a useful method for the preparation of diverse 1,1-diaryl methanes and their derivatives.

7.4   Experimental Section and Compound Data

7.4.1   General Information

All the reactions were carried out in oven-dried Schlenk tubes under Argon atmosphere (purity ≥ 99.999%). The mesitylene solvent was bought from TCI and used without further purification. All aryl halides were bought from Alfa Aesar or TCI and used as is. Nitro phenylacetic acids were purchased from TCI. Allyl palladium chloride dimer was purchased from Sigma-Aldrich and Sinocompound. All phosphine ligands were bought from Sinocompound, Sigma-Aldrich, Strem or Alfa Aesar and used as is. All the other reagents and solvents mentioned in this text were bought from Sinopharm Chemical Reagent Co. Ltd or Alfa Aesar and purified when necessary.

^{1}H-NMR, ^{13}C-NMR spectra were recorded on a Bruker Avance 400 spectrometer at ambient temperature in CDCl$_3$ unless otherwise noted. Data for ^{1}H-NMR are reported as follows: chemical shift (δ ppm), multiplicity, integration, and coupling constant (Hz). Data for ^{13}C-NMR are reported in terms of chemical shift (δ ppm), multiplicity, and coupling constant (Hz). Gas chromatographic (GC) analysis was acquired on a Shimadzu GC-2014 Series GC System equipped with a flame-ionization detector. GC-MS analysis was performed on Thermo Scientific AS 3000 Series GC-MS System. HRMS analysis was performed on Finnigan LCQ advantage Max Series MS System. Organic solutions were concentrated under reduced pressure on a Buchi rotary evaporator. Flash column chromatographic purification of products was accomplished using forced-flow chromatography on Silica Gel (200–300 mesh).

7.4.2   General Procedures

The decarboxylative coupling of potassium nitrophenyl acetates with phenylhalides:

A 10 mL oven-dried Schlenk tube was charged with Pd source (see Table 7.4), phosphine ligand (see Table 7.4), and potassium nitrophenyl acetates (see Table 7.4, 0.30 mmol). The tube was evacuated and filled with argon (this procedure was repeated three times). Then phenylhalide (0.25 mmol) and mesitylene (0.5 mL) were added with a syringe under a counter flow of argon. The tube was sealed with a screw cap, stirred at room temperature for 1 min, and connected to the Schlenk line that was full of argon, stirred in a preheated oil bath (140 °C) for 5 h. Upon completion of the reaction, the mixture was cooled to room temperature and diluted with ethyl acetate, the yields were determined by gas chromatography using benzophenone as internal standard.

Some structures of the ligands mentioned in Table 7.4 are shown in Fig. 7.1.

Control Experiments

We performed the radical trap experiments (see Scheme 7.3). The results rule out the involvement of a radical mechanism.

Table 7.4 Pd-catalyzed decarboxylative cross-coupling of potassium nitrophenyl acetates with phenyl halide[a]

Entry	X	o, m, p	Pd source	Ligand	Yield %[a]	
					1	2
1	Cl	o	[PdCl(allyl)]$_2$	P(Ad)$_2^n$Bu	83	18
2	Cl	o	[PdCl(allyl)]$_2$	P(^{t}Bu)$_3$	<5	61
3	Cl	o	[PdCl(allyl)]$_2$	X-Phos	91	16
4	Cl	o	[PdCl(allyl)]$_2$	t-BuX-Phos	<5	74
5	Cl	o	[PdCl(allyl)]$_2$	S-Phos	85	16
6	Cl	o	[PdCl(allyl)]$_2$	Dave-Phos	83	17
7	Cl	o	[PdCl(allyl)]$_2$	DCyPF	51	43
8	Cl	o	[PdCl(allyl)]$_2$	Brett-Phos	56	36
9	Cl	p	[PdCl(allyl)]$_2$	X-Phos	89	16
10	Cl	m	[PdCl(allyl)]$_2$	X-Phos	n.r.	<5
11	Br	o	[PdCl(allyl)]$_2$	DPPF	80	22
12	Br	o	[PdCl(allyl)]$_2$	BINAP	86	18
13	Br	o	[PdCl(allyl)]$_2$	Xant-Phos	81	23
14	Br	o	[PdCl(allyl)]$_2$	DPE-Phos	53	46
15	Br	o	[PdCl(allyl)]$_2$	DPPP	37	48
16	Br	p	[PdCl(allyl)]$_2$	BINAP	88	17
17[b]	Cl	o	Pd(OAc)$_2$	X-Phos	7	73
18	Cl	o	Pd$_2$(dba)$_3$	X-Phos	87	21

[a]All the reactions were carried out in 0.25 mmol scale in 0.5 mL mesitylene. GC yields, average of two runs, using benzophenone as internal standard. [b]4% mol Pd source

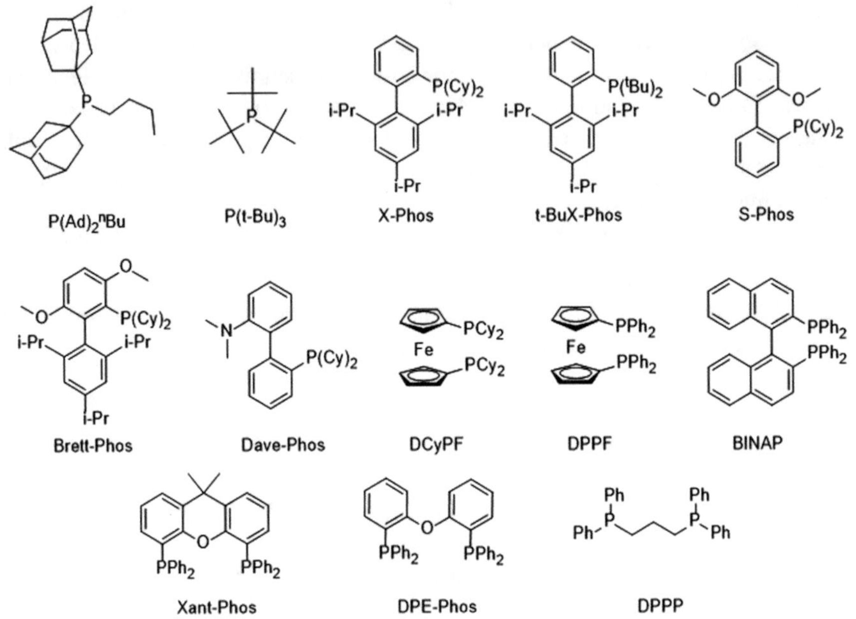

P(Ad)$_2^n$Bu   P(t-Bu)$_3$ X-Phos   t-BuX-Phos   S-Phos

Brett-Phos   Dave-Phos   DCyPF   DPPF   BINAP

Xant-Phos   DPE-Phos   DPPP

Fig. 7.1 Some structures of the ligands mentioned in Table 7.4

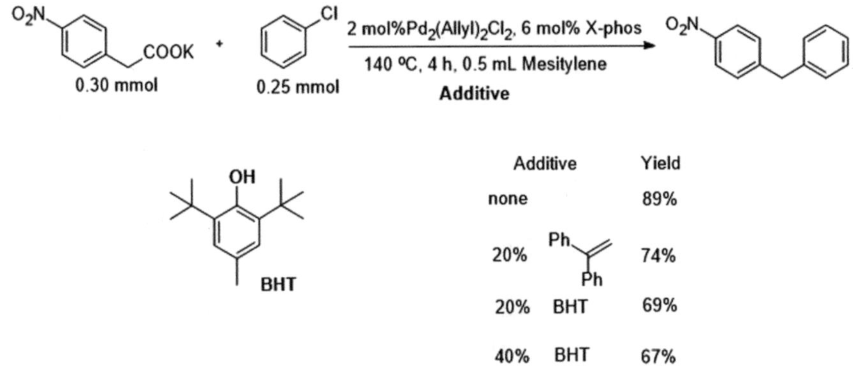

Additive		Yield
none		89%
20%	Ph⧸Ph	74%
20%	BHT	69%
40%	BHT	67%

Scheme 7.3 Radical trap experiments

Our control experiments showed that in the absence of Pd, the reaction does not take place (Scheme 7.4). Specifically, the potassium salt does not decarboxylate at 140 °C in the absence of Pd. It must be noted that the aryl-acetic acid can undergo the protodecarboxylation reaction, but the potassium salt does not decarboxylate without a metal catalyst.

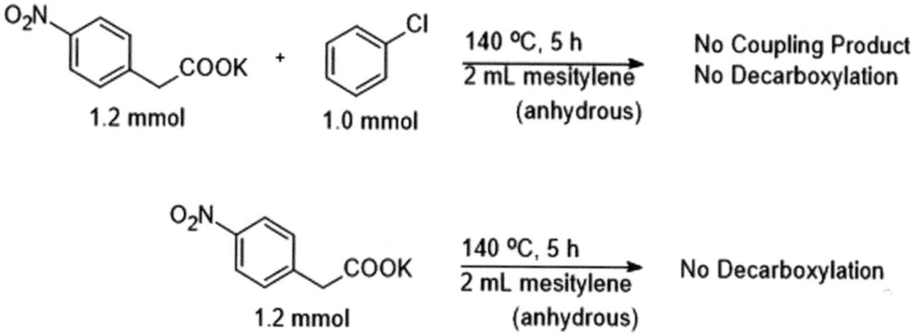

Scheme 7.4 Control experiments oC in the absence of Pd

7.4.3 *Characterization of the Products*

Compound name 1-benzyl-2-nitrobenzene

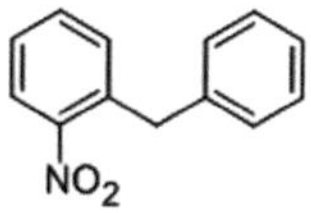

94 mg (88%) of a yellow oil was obtained. This compound is known. Reference: Lu, F.; Chi, S.; Kim, D. H.; Han, K. H.; Kuntz, I. D.; Guy, R. K. J. Comb. Chem. 2006, 8, 315. ^{1}H NMR (400 MHz, CDCl$_3$) δ 7.90 (dd, J = 8.1, 1.3 Hz, 1H), 7.48 (td, J = 7.6, 1.4 Hz, 1H), 7.34 (td, J = 8.1, 1.4 Hz, 1H), 7.30–7.23 (m, 3H), 7.21 (d, J = 6.8 Hz, 1H), 7.16–7.11 (m, 2H), 4.29 (s, 2H). ^{13}C NMR (100 MHz, CDCl$_3$) δ 149.29, 138.62, 135.62, 132.85, 132.32, 128.93, 128.54, 127.31, 126.52, 124.65, 38.20.

Compound name 4-(2-nitrobenzyl)phenyl 4-methylbenzenesulfonate

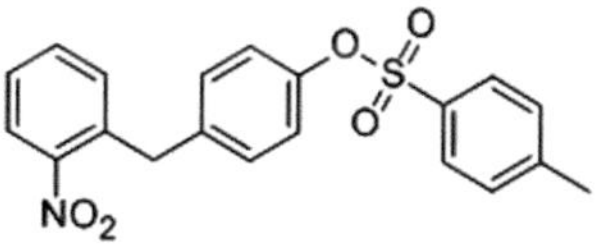

153 mg (80%) of a yellow oil was obtained. This compound is new. ^{1}H NMR (400 MHz, CDCl$_3$) δ 7.90 (dd, J = 8.1, 1.2 Hz, 1H), 7.65 (d, J = 8.3 Hz, 2H), 7.52 (td, J = 7.6, 1.3 Hz, 1H), 7.41–7.34 (m, 1H), 7.32–7.22 (m, 3H), 7.04 (d, J = 8.7 Hz, 2H), 6.91–6.83 (m, 2H), 4.24 (s, 2H), 2.42 (s, 3H). ^{13}C NMR (100 MHz, CDCl$_3$) δ 149.01, 148.07, 145.30, 137.68, 134.75, 133.04, 132.36, 132.09, 129.86, 129.63, 128.28, 127.63, 124.75, 122.25, 37.63, 21.52 . HRMS calcd for C$_{20}$H$_{17}$NO$_5$ S (M+) 383.0822; found: 383.0819.

Compound name 4-(2-nitrobenzyl)benzonitrile

75 mg (63%) of a yellow solid was obtained. This compound is known. Reference: Lu, F.; Chi, S.-W.; Kim, D.-H.; Han, K.-H.; Kuntz, I. D.; Guy, R. K. J. Comb. Chem. 2006, 8, 315. ^{1}H NMR (400 MHz, CDCl$_3$) δ 8.01 (dd, J = 8.2, 1.3 Hz, 1H), 7.62–7.55 (m, 3H), 7.48–7.42 (m, 1H), 7.30 (dd, J = 7.7, 1.0 Hz, 1H), 7.27–7.23 (m, 2H), 4.37 (s, 2H). ^{13}C NMR (100 MHz, CDCl$_3$) δ 149.10, 144.30, 133.91, 133.38, 132.63, 132.35, 129.48, 128.18, 125.21, 118.71, 110.57, 38.70 . HRMS calcd for C$_{14}$H$_{10}$N$_2$O$_2$ (M+) 238.0737; found: 238.0733.

Compound name ethyl 4-(2-nitrobenzyl)benzoate

97 mg (68%) of a yellow oil was obtained. This compound is new. ^{1}H NMR (400 MHz, CDCl$_3$) δ 7.99–7.94 (m, 3H), 7.54 (td, J = 7.6, 1.4 Hz, 1H), 7.44–7.38 (m, 1H), 7.27 (dd, J = 7.8, 1.1 Hz, 1H), 7.21 (d, J = 8.5 Hz, 2H), 4.39–4.32 (m, 4H), 1.38 (t, J = 7.1 Hz, 3H). ^{13}C NMR (100 MHz, CDCl$_3$) δ 166.35, 149.23, 143.87, 134.76, 133.11, 132.46, 129.85, 128.92, 128.83, 127.75, 124.95, 60.85, 38.43, 14.29 . HRMS calcd for C$_{16}$H$_{15}$NO$_4$ (M+) 285.0996; found: 285.0994.

Compound name 1-nitro-2-(4-(trifluoromethyl)benzyl)benzene

107 mg (76%) of a yellow oil was obtained. This compound is new. ^{1}H NMR (400 MHz, CDCl$_3$) δ 7.96 (dd, J = 8.1, 1.3 Hz, 1H), 7.56 (dd, J = 7.5, 1.3 Hz, 1H), 7.52 (d, J = 7.5 Hz, 2H), 7.44–7.37 (m, 1H), 7.29 (dd, J = 7.7, 0.9 Hz, 1H), 7.26 (d, J = 8.0 Hz, 2H), 4.36 (s, 2H). ^{13}C NMR (100 MHz, CDCl$_3$) δ 149.15, 142.88, 134.48, 133.23, 132.31, 129.10, 128.84 (q, J_F = 32.0 Hz), 127.88, 125.44 (q, J_F = 4.0 Hz), 124.99, 121.44 (q, J_F = 254.5 Hz), 38.25. HRMS calcd for C$_{14}$H$_{10}$F$_3$NO$_2$ (M+) 281.0658; found: 281.0661.

Compound name N,N-dimethyl-4-(2-nitrobenzyl)aniline

110 mg (86%) of an orange solid was obtained. This compound is new. ^{1}H NMR (400 MHz, CDCl$_3$) δ 7.87 (dd, J = 8.1, 1.2 Hz, 1H), 7.46 (td, J = 7.6, 1.3 Hz, 1H), 7.35–7.28 (m, 1H), 7.28–7.23 (m, 1H), 7.02 (d, J = 8.7 Hz, 2H), 6.70–6.64 (m, 2H), 4.19 (s, 2H), 2.91 (s, 6H). ^{13}C NMR (100 MHz, CDCl$_3$) δ 149.34, 136.75, 132.70, 132.11, 129.73, 126.94, 126.45, 124.47, 112.83, 40.62, 37.16 . HRMS calcd for C$_{15}$H$_{16}$N$_2$O$_2$ (M+) 256.1206; found: 256.1201.

Compound name 1-nitro-2-(4-vinylbenzyl)benzene

98 mg (82%) of a yellow oil was obtained. This compound is new. ^{1}H NMR (400 MHz, CDCl$_3$) δ 7.91 (dd, J = 8.1, 1.3 Hz, 1H), 7.49 (td, J = 7.6, 1.3 Hz, 1H), 7.39–7.23 (m, 4H), 7.10 (d, J = 8.1 Hz, 2H), 6.67 (dd, J = 17.6, 10.9 Hz, 1H), 5.70 (dd, J = 17.6, 0.8 Hz, 1H), 5.20 (dd, J = 10.9, 0.7 Hz, 1H), 4.28 (s, 2H). ^{13}C NMR (100 MHz, CDCl$_3$) δ 149.28, 138.25, 136.38, 135.97, 135.57, 132.91, 132.30, 129.12, 127.37, 126.42, 124.72, 113.56, 37.99. HRMS calcd for C$_{15}$H$_{13}$NO$_2$ (M+) 239.0941; found: 239.0936.

Compound name trimethyl(4-(2-nitrobenzyl)phenyl)silane

127 mg (89%) of a brownish oil was obtained. This compound is new. ^{1}H NMR (400 MHz, CDCl$_3$) δ 7.92 (dd, J = 8.1, 1.3 Hz, 1H), 7.50 (td, J = 7.6, 1.3 Hz, 1H), 7.44 (d, J = 8.0 Hz, 2H), 7.40–7.33 (m, 1H), 7.28 (dd, J = 7.7, 0.8 Hz, 1H), 7.14 (d, J = 7.9 Hz, 2H), 4.29 (s, 2H), 0.24 (s, 9H). ^{13}C NMR (100 MHz, CDCl$_3$) δ 149.33, 139.21, 138.43, 135.60, 133.63, 132.91, 132.43, 128.34, 127.35, 124.72, 38.23, −1.14 . HRMS calcd for C$_{16}$H$_{19}$NO$_2$Si (M+) 285.1180; found: 285.1183.

Compound name 2-methyl-6-(2-nitrobenzyl)benzo[d]thiazole

131 mg (92%) of a brownish solid was obtained. This compound is new. ^{1}H NMR (400 MHz, CDCl$_3$) δ 7.94 (dd, J = 8.1, 1.3 Hz, 1H), 7.71 (d, J = 8.2 Hz, 1H), 7.67 (d, J = 0.9 Hz, 1H), 7.50 (td, J = 7.6, 1.3 Hz, 1H), 7.40–7.33 (m, 1H), 7.30 (dd, J = 7.7, 0.9 Hz, 1H), 7.18 (dd, J = 8.2, 1.6 Hz, 1H), 4.43 (s, 2H), 2.79 (s, 3H). ^{13}C NMR (100 MHz, CDCl$_3$) δ 167.41, 153.70, 149.13, 136.90, 135.35, 133.77, 132.96, 132.44, 127.47, 125.82, 124.79, 122.34, 121.28, 38.14, 20.00 . HRMS calcd for C$_{15}$H$_{12}$N$_2$O$_2$S (M+) 284.0614; found: 284.0608.

Compound name tert-butyl 5-(4-nitrobenzyl)-1H-indole-1-carboxylate

111 mg (63%) of a yellow oil was obtained. This compound is new. ^{1}H NMR (400 MHz, CDCl$_3$) δ 7.99 (d, J = 8.0 Hz, 3H), 7.49 (d, J = 3.6 Hz, 1H), 7.23 (s, 1H), 7.21 (d, J = 8.7 Hz, 2H), 7.00 (dd, J = 8.5, 1.6 Hz, 1H), 6.40 (d, J = 3.6 Hz, 1H), 4.03 (s, 2H), 1.56 (s, 9H). ^{13}C NMR (100 MHz, CDCl$_3$) δ 149.55, 149.42, 146.33, 133.97, 133.44, 130.95, 129.47, 126.37, 125.19, 123.57, 121.01, 115.34, 106.95, 83.66, 41.51, 28.07 . HRMS calcd for C$_{20}$H$_{20}$N$_2$O$_4$ (M+) 352.1418; found: 352.1414.

Compound name 1-nitro-4-(1-phenylethyl)benzene

83 mg (73%) of a yellow oil was obtained. This compound is known. Reference: Inoh, J.-I.; Satoh, T.; Pivsa-Art, S.; Miura, M.; Nomura, M. Tetrahedron Lett. 1998, 39, 4673. ^{1}H NMR (400 MHz, CDCl$_3$): δ 8.13 (d, J = 9.2 Hz, 2H), 7.36 (d, J = 8.4 Hz, 2H), 7.31 (t, J = 7.2 Hz, 2H), 7.25–7.16 (m, 3H), 4.25 (q, J = 7.2 Hz, 1H), 1.67 (d, J = 7.2 Hz, 3H). ^{13}C NMR (100 MHz, CDCl$_3$): δ 154.17, 146.53, 144.66, 128.85, 128.57, 127.68, 126.84, 123.83, 44.88, 21.63. HRMS calcd for C$_{14}$H$_{13}$NO$_2$ (M+) 227.0941; found: 227.0931.

Compound name 3-(2-nitrophenyl)-1-phenyl-3-(4-(trifluoromethyl)phenyl)propan-1-one

colorless oil. ^{1}H NMR (400 MHz, CDCl$_3$) δ 7.96–7.91 (m, 2H), 7.87–7.81 (m, 1H), 7.57–7.49 (m, 4H), 7.47–7.40 (m, 4H), 7.39–7.33 (m, 2H), 5.53 (t, J = 7.3 Hz, 1H), 3.84 (dd, J = 17.0, 6.8 Hz, 1H), 3.78 (dd, J = 17.0, 6.4 Hz, 1H). HRMS calcd for C$_{22}$H$_{16}$F$_3$NO$_3$ (M+) 399.1082; found: 399.1077.

References

1. Weaver, J. D., Recio, III, A., Grenning, A. J., & Tunge, J. A. (2011). *Chemical Reviews, 111*, 1846–1913.
2. Myers, A. G., Tanaka, D., & Mannion, M. R. (2002). *Journal of the American Chemical Society, 124*, 11250–11251.

3. Forgione, P., Brochu, M. C., St-Onge, M., Thesen, K. H., Bailey, M. D., & Bilodeau, F. (2006). *Journal of the American Chemical Society, 128,* 11350–11351.

4. Goossen, L. J., Deng, G., & Levy, L. M. (2006). *Science, 313,* 662–664.

5. Maehara, A., Tsurugi, H., Satoh, T., & Miura, M. (2008). *Organic Letters, 10,* 1159–1162.

6. Rayabarapu, D. K., & Tunge, J. A. (2005). *Journal of the American Chemical Society, 127,* 13510–13511.

7. Trost, B. M., Xu, J., & Schmidt, T. (2008). *Journal of the American Chemical Society, 130,* 11852–11853.

8. Mohr, J. T., Nishimata, T., Behenna, D. C., & Stoltz, B. M. (2006). *Journal of the American Chemical Society, 128,* 11348–11349.

9. Nakamura, M., Hajra, A., Endo, A. H., & Nakamura, E. (2010). *Angewandte Chemie International Edition, 2005, 44,* 7248–7251.

10. Bi, H.-P., Zhao, L., Liang, Y.-M., Li, & C.-J. (2009). *Angewandte Chemie International Edition, 48,* 792–795.

11. Shang, R., Yang, Z. W., Wang, Y., Zhang, S.-L., & Liu, L. (2010). *Journal of the American Chemical Society, 132,* 14391–14393.

12. Shang, R., Ji, D. S., Chu, L., Fu, Y., & Liu, L. (2011). *Angewandte Chemie International Edition, 50,* 4470–4474.

13. *The Nitro Group in Organic Synthesis, Ono, N., Ed., Wiley-VCH: New York.* **2001**.

14. Inoh, J. I., Satoh, T., Pivsa-Art, S., Miura, M., & Nomura, M. (1998). *Tetrahedron Letters, 39,* 4673–4676.

15. Base-mediated decarboxylation of nitrophenyl acetic acids can take place to produce alkylnitrobenzenes; see: Bull, D. J., Fray, M. J., Mackenny, M. C., Malloy, & K. A. (1996). *Synlett, 7,* 647–648. However, the potassium salt of 4-nitrophenyl acetic acid does not decarboxylate at 140 °C in the absence of palladium catalyst.

16. Prasad, G., Hanna, P. E., Noland, W. E., & Venkatraman, S. (1991). *The Journal of Organic Chemistry, 56,* 7188–7190.

17. Parisien, M., Valette, D., & Fagnou, K. (2005). *Journal of Organic Chemistry, 70,* 7578–7584.

18. We also tested the potassium salts of 4-cyanophenyl acetate, 4-trifluoromethylphenyl acetate, and pentafluorophenyl acetate. These substrates do not provide the cross-coupling products.

Chapter 8
Palladium-Catalyzed Decarboxylative Benzylation of α-Cyano Aliphatic Carboxylate Salts with Benzyl Electrophiles

Abstract The palladium-catalyzed decarboxylative benzylation of α-cyano aliphatic carboxylate salts with benzyl electrophiles was discovered. This reaction exhibits good functional group compatibility and proceeds under relatively mild conditions. A diverse range of quaternary, tertiary, and secondary β-aryl nitriles can be conveniently prepared by this method with good to excellent yield, and many of them are difficult to be synthesized by strong base-mediated nucleophilic substitution.

8.1 Introduction

Transition metal-catalyzed decarboxylative cross-coupling has represented a useful strategy for carbon–carbon bond construction in modern synthetic organic chemistry [1]. Inspired by the seminal work of Myers [2] and Gooßen [3] catalytic decarboxylative coupling reactions of aromatic [4], alkenyl [5], and alkynyl [6] carboxylic acids have been intensively studied. More recently, our group, Studer, and Kwong et al. developed a palladium-catalyzed decarboxylative arylation of some aliphatic carboxylate salts with aryl halides [7]. Tunge et al. reported excellent studies on intramolecular decarboxylative allylations of some activated carboxylate allylic esters [8], and an intramolecular decarboxylative benzylation of β-keto esters, but the benzyl groups participating in this chemistry were limited to benzyls with extend conjugations (Fig. 8.1) [9]. To the best of our knowledge, intermolecular decarboxylative benzylation of aliphatic carboxylates has not been explored. Because catalytic reactions involving π-benzyl-Pd intermediates are

© Springer Nature Singapore Pte Ltd. 2017
R. Shang, *New Carbon–Carbon Coupling Reactions Based on Decarboxylation and Iron-Catalyzed C–H Activation*, Springer Theses,
DOI 10.1007/978-981-10-3193-9_8

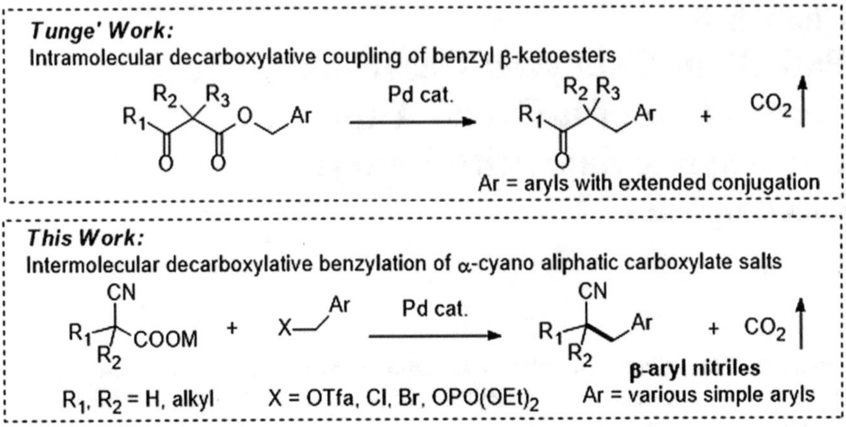

Fig. 8.1 Decarboxylative benzylation of Csp^3-COOR

important and have drawn broad attention [10], after developing the decarboxyla-tive arylation of activated alkyl carboxylate, we were interested in the inter-molecular decarboxylative benzylation of acyano aliphatic carboxylates.

Traditionally, the β-aryl nitrile structures were constructed by strong base-mediated nucleophilic substitution of nitrile's α–C–H with benzyl halides [11]. Using strong bases largely limits the substrate scope and ruins base-sensitive functional groups. In addition, for acetonitrile and primary nitriles, monobenzyla-tion is often complicated due to the formation of multiply benzylated by-products. Benzylation of ethyl cyanocaetate followed by hydrolysis and decarboxylation can be used for the synthesis of some secondary and tertiary β-aryl nitriles efficiently, but is unable to give the quaternary ones [12]. Due to the importance of func-tionalized nitriles in both synthetic and medicinal chemistry [13], we now report an intermolecular Pd-catalyzed decarboxylative benzylation reaction of α-cyano ali-phatic carboxylate salts with benzyl electrophiles [–Cl, –OTfa, –OPO(OEt)$_2$]. This reaction provides an efficient approach to construct a diverse range of functionalized quaternary, tertiary and secondary β-aryl nitriles, and has a wide substrate scope of carboxylate salts with benzyl electrophiles. The reaction also represents the first example of intermolecular decarboxylative benzylation of aliphatic carboxylates, and expands the synthetic utility of the decarboxylative coupling methodology.

8.2 Results and Disscussion

8.2.1 Investigation of the Reaction Conditions

We initially chose benzyl chloride and potassium 2-cyano-2-methylpropanoate as the model substrates for screening the decarboxylative benzylation conditions

Table 8.1 Decarboxylative benzylation under various conditions

$$Ph\frown Cl \;(0.25\ mmol) \;+\; MOOC\overset{CN}{\diagup} \;(0.375\ mmol,\ M = Li,\ Na,\ K) \xrightarrow[\substack{0.5\ mL\ Solvent \\ T,\ 5\ h}]{\substack{2\ mol\%\ [Pd(allyl)Cl]_2 \\ 6\ mol\%\ Ligand}} Bn\overset{CN}{\diagup}\ \mathbf{1} \;+\; BnOOC\overset{CN}{\diagup}\ \mathbf{2}$$

Entry	M	T (°C)	Ligand	Solvent	Yield 1	Yield 2
1	K	140	Xant-Phos	Mesitylene	8	53
2	K	140	BINAP	Mesitylene	54	9
3	K	140	DPPF	Mesitylene	80	Trace
4	K	140	DPE-Phos	Mesitylene	56	Trace
5	K	140	DPPPent	Mesitylene	71	–
6	K	140	CataCXium A	Mesitylene	18	–
7	K	140	Ph-Dave-Phos	Mesitylene	78	
8	K	140	t-BuX-Phos	Mesitylene	32	–
9	K	140	Ru-Phos	Mesitylene	95	–
10	K	140	X-Phos	Mesitylene	94	–
11	K	140	S-Phos	Mesitylene	98	–
12	K	140	IPr·HCl	Mesitylene	74	–
13	Na	140	S-Phos	Mesitylene	94	–
14[a]	Li	140	S-Phos	Mesitylene	Trace	–
15	K	140	S-Phos	Diglyme	95	–
16	K	140	S-Phos	DMA	51	–
17	K	140	S-Phos	NMP	57	–
18	K	140	S-Phos	DMSO	20	6
19	K	110	S-Phos	Mesitylene	95	–
20	K	100	S-Phos	Mesitylene	95	–
21	K	110	S-Phos	Toluene'	95	–

[a]Recovery of benzyl chloride in 90% yield

(Table 8.1). Firstly, we tested the catalyst composed of [Pd(allyl)Cl]$_2$ and Xant-Phos, which have been shown to be the best choice for our previously reported decarboxylative arylation process, but the formation of benzyl ester was observed (entry 1). Further experiments showed that the ratio of the decarboxylative benzylation product and benzyl ester was largely affected by the choice of the ligand. By screening the ligands, we found that when 1'-Bis(diphenylphosphino) ferrocene (dppf) was used as the ligand, the decarboxylative benzylation product was obtained in 80% yield, accompanied with only a trace amount of the benzyl ester by-product (entry 3). Inspired by this result, we screened the ligands extensively, and found that Buchwald phosphines were the best choice. When S-Phos was chosen as the ligand, the desired transformation took place efficiently (98% yield) without detection of the benzyl ester. X-Phos and Ru-Phos, which posses the

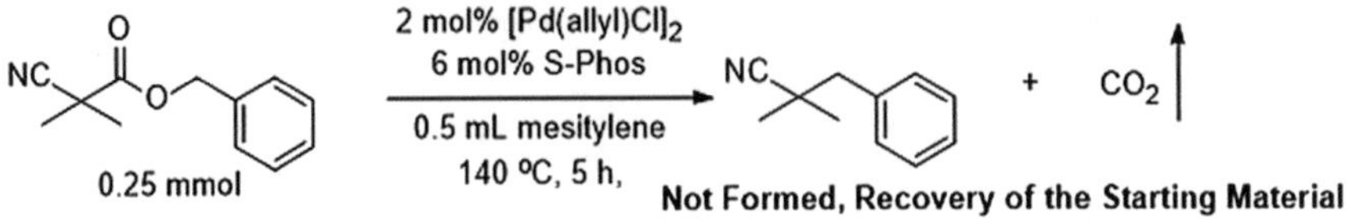

Scheme 8.1 Control experiment for intramolecular ester decarboxylative benzylation

Table 8.2 Survey of the benzyl electrophiles

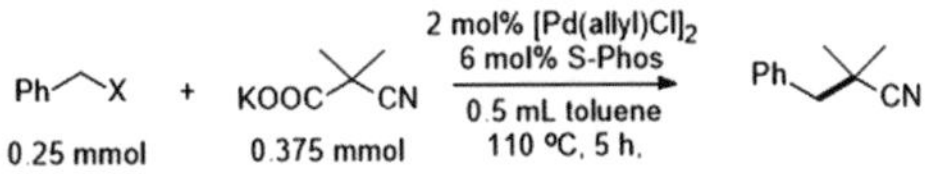

Entry	X	Yield (GC) %	Entry	X	Yield (GC) %
1	Cl	95	5	OTFA	86
2	Br	73	6	$OPO(OE)_2$	93
3	OAc	0	7	OB_{oc}	12
4	OTs	64	8	OCO(OEt)	3

Reproduced from Adv. Syn. Catal. 2012, 354, 2465 by permission of John Wiley & Sons Ltd.

same structure with S-Phos, were also the effective ligands for the reaction. We found that sodium salts can also serve as good substrates (entry 13), but the lithium salts failed to give any desired product (entry 14). Screening of the solvents revealed that the reaction had a strong solvent effect, and non-polar arene solvents were good choices. When diglyme was used as the solvent, the product was obtained in 95% yield (entry 15). Polar solvents, such as amides and sulfones, were less effective (Entrise 16, 17, 18). To our delight, the reaction can take place under relatively mild conditions (entry 20) and can proceed well in refluxing toluene (entry 21). Finally, we tested the intramolecular decarboxylative benzylation of benzyl 2-cyano-2-methylpropanoate under the optimized conditions but failed to obtain the benzylated product (Scheme 8.1) [14].

After finding the optimal conditions, we tested the scope of the leaving groups of the benzyl electrophiles (Table 8.2). Experiment results showed that benzyl bromide, chloride, tosylate, trifluoroacetate, and diethyl phosphate could serve as suitable electrophiles under the optimal conditions. Benzyl carbonates such as –OBoc and –OCO(OEt)$_2$ gave low yields, while benzyl acetate could not be activated under the optimized conditions with total recovery of the starting material.

8.2.2 Study of the Substrate Scope

Next, we evaluated the scope of the various substituted benzyl electrophiles (Table 8.3). From the perspective of convenience, the stable benzyl chlorides can

Table 8.3 Decarboxylative benzylation of 2-cyano-2-methylpropanoate salts[a]

[a]Reactions were carried out on a 0.5 mmol scale. Yields are isolated yields based on the quantity of benzyl electrophiles

be easily prepared from the corresponding benzyl alcohol. For this reason, we decided to evaluate the scope of substituted benzyl electrophiles mainly using benzyl chlorides. To our delight, a variety of functionalized benzyl electrophiles could be used for the benzylation of potassium 2-cyano-2-methylpropanoate under the optimized conditions achieving good to excellent yields. The electrophiles carrying substituents on *ortho* (3a, 3e), *meta-* (3b, 3d), and *para-* (3c) positions of the benzyl ring are all amenable substrates. The *para*-methyl substitution on the benzyl ring results in a relatively low yield (3c) [15]. Benzyl electrophiles carrying either electron-withdrawing or electron-donating functional groups including alkyl (3a–3d), aryl (3e, 3j), alkoxy (3r, 3s, 3t), aryloxy (3g), fluoro (3h), chloro (3i), trifluoromethyl (3n), ester (3 m), electron-rich and electron-deficient C=C double bond (3o, 3p), and amine (3q) can all react with good yields.

3-Chloromethylthiophene can react and the product (3 l) was obtained in 87% yield. Pyridine-3-methanol-derived phosphate can also react, albeit in relatively lower yield (3k, 30%). It is important to note that alkyl pinacol boronate and alkyl chloride functionality can be well tolerated in the present transformation (3r, 3 s), which can be further transformed via other alkyl cross-coupling reactions [16]. The presence of base-sensitive, enolizable acidic C–H bond on the benzyl electrophiles can be well tolerated (3t), and the product can be further modified via reactions on enolizable C–H bond [17]. Note that this product (3t) is difficult to synthesize via strong base-mediated deprotonation/nucleophilic substitution pathways due to the presence of the acidic C–H moiety. Benzyl electrophiles with extended conjugations can also react, such as 1-chloromethyl naphthalene (3u) and 2-naphthyl methanol-derived phosphate (3v). Unfortunately, α-substituted benzyl electrophiles failed to react under the present catalytic conditions (3w, 3x).

We next explored the scope of the 2-cyano aliphatic carboxylate salts (Table 8.4). Potassium cyanoacetate can react well with benzyl electrophiles to form β-arylpropanenitriles cleanly without forming any dibenzylated by-product (4a, 4b) [18]. Tertiary α-cyano aliphatic carboxylate such as potassium 2-cyanopropanoate (4c, 4d), potassium 2-cyano-4-phenylbutanoate (4e), potassium 2-cyanohexanoate (4f, 4 g), and potassium 2-cyanooctanoate (4i, 4j) were all amenable to the decarboxylative benzylation reaction, and the tertiary β-aryl nitrile products were obtained in moderate to good yields. Potassium 2-cyano-3-methylbutanoate gave a low yield (28%, 4h). This is possibly due to the steric hindrance of the adjacent methyl group. Next, we examined the quaternary salt substrates particularly. The benzylation of this kind of substrate can be used to synthesize quaternary α-cyanide. Sodium 2-cyano-2-methylhexanoate can react well with 3-methoxybenzyl chlorides to afford the corresponding product (4l) with 73% yield. Quaternary α-cyano aliphatic carboxylate salts carrying aryl substitution on the aliphatic chain also serve as suitable substrates. The aryl substituents on aliphatic chain expand the scope of quaternary α-cyano aliphatic carboxylate salts (4q, 4r, 4 s). Some other substituents on the aliphatic chain of the α-cyano aliphatic carboxylates were also investigated. The results show that pinacol boronate functionality on the aliphatic chain can be well tolerated. This example illustrates that the alkyl decarboxylative benzylation shows selectivity toward traditional alkyl Suzuki coupling thus giving the further possibility of modification via transformations of alkyl boronate (4k) [16]. α-Cyclic and oxygen containing cyclic carboxylate salts also serve as good substrates (4m, 4n, 4t). It is important to note that cyano and ketone functionalities on the aliphatic chain were well tolerated (4o, 4p, 4u). The decarboxylative benzylation took place regiospecifically replacing the carboxylate moiety to form a quaternary carbon center without any benzylation on enolizable C–H bond. These examples highlight the synthetic utility of the Pd-catalyzed decarboxylative benzylation processes.

During our optimization study for the coupling of benzyl chloride with potassium 2-cyano-2-methylpropanoate, we observed a by-product that showed equal molecular weight with the desired product by GC-MS analysis. NMR analysis indicated that the by-product is 2-methyl-2-(p-tolyl)propanenitrile. Further

Table 8.4 Decarboxylative benzylation of various α-cyano aliphatic carboxylate salts[a]

$$R_1 \underset{R_2}{\overset{CN}{\mid}} COOK \;+\; X\text{-}Ar \xrightarrow[\substack{110\ ^\circ C,\ 1\ mL\ toluene, \\ 5\text{-}12\ h}]{\substack{2\ mol\%\ [Pd(allyl)Cl]_2 \\ 6\ mol\%\ S\text{-}Phos}} R_1 \underset{R_2}{\overset{CN}{\mid}} Ar \;+\; CO_2 \uparrow$$

0.5-0.75 mmol   0.5 mmol   X = Cl, OPO(OEt)$_2$

X-phos
NC
4a, - Cl, 85%
-OPO(OEt)$_2$, 72%

X-phos
NC, -O-Ph
-Cl, 4b, 87%

X-phos CN
4c, - Cl, 78%
-OPO(OEt)$_2$, 70%

CN ...Ph
- Cl, 4d, 78%

CN, Ph, COOEt
- Cl, 4e, 65%

CN, O-Ph
-Cl, 4f, 66%   X-phos

CN, COO^{n}Bu
- Cl, 4g, 43%

CN, Ph
- Cl, 4h, 28%

CN, Cl
- Cl, 4i, 72%

CN, CF$_3$
- Cl, 4j, 45%

Bpin, CN, COOEt
Sodium Salt
- Cl, 4k, 40%

CN, O-
Sodium Salt
- Cl, 4l, 73%

NC, (O ring)
- Cl, 4m, 78%

NC, (O ring), N-O
- Cl, 4n, 62%

O, CN, Ph
- Cl, 4o, 88%

O, CN, Ph, COO^{n}Bu
- Cl, 4p, 73%

Ph, CN, O-
- Cl, 4q, 48%

Ph, CN, Ph, O-
- Cl, 4r, 52%

CN, Ph, N-O
- Cl, 4s, 65%

NC (cyclopentyl)
- Cl, 4t, 80%

NC, Sodium Salt, CN, COOEt, O-
- Cl, 4u, 66%

Reproduced from Adv. Syn. Catal. 2012, 354, 2465 by permission of John Wiley & Sons Ltd.

[a]Reactions were carried out on a 0.5 mmol scale. Isolated yields based on the quantity of benzyl electrophiles

investigation showed the ratio of the by-product is largely affected by the choice of the solvent (Scheme 8.2). Although the detailed mechanism and the solvent-induced selectivity for the formation of the arylated product are not clear, the by-product formation process may involve a dearomatization of the Pd-benzyl intermediate according to the literature [19].

We tried the decarboxylative alkylation of alkyl bromides with potassium 2-cyano-2-methylpropanoate, however, no alkylation product was determined under the optimal conditions. NMR analysis showed that the attempted coupling reaction led to the formation of alkyl ester in high yield (Scheme 8.3). Interestingly,

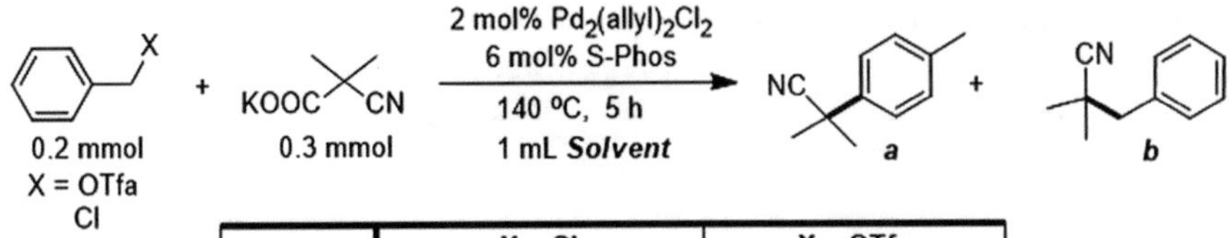

Scheme 8.2 Solvent effect on reaction selectivity

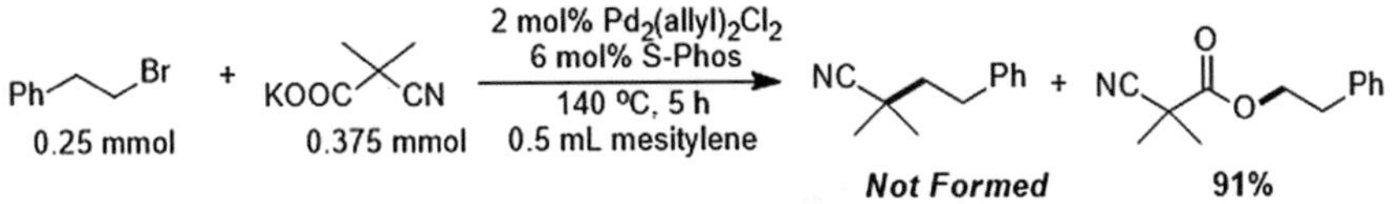

Solvent	X = Cl		X = OTfa	
	a (%)	b (%)	a (%)	b (%)
mesitylene	3	95	2	95
NMP	50	45	34	53

Scheme 8.3 Tentative decarboxylative coupling of primary alkyl bromides

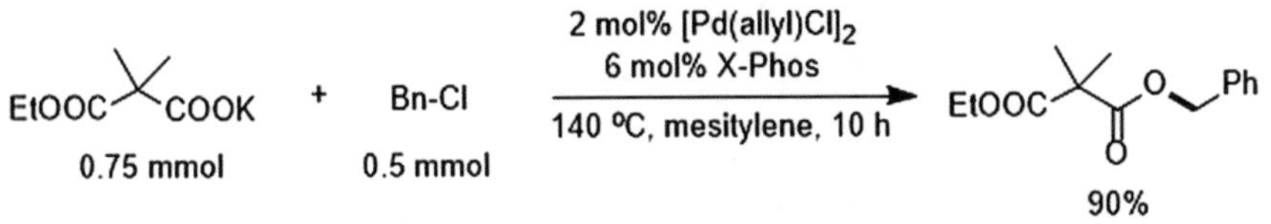

Scheme 8.4 Tentative decarboxylative benzylation of potassium 3-ethoxy-2,2-dimethyl-3-oxopropanoate

when the above reaction was carried out in the absence of palladium catalyst, no alkyl ester was formed. This observation indicates that the alkyl ester formation is not a simple S_N2 type reaction, but is catalyzed by palladium under the reaction condition. We also tried the decarboxylative benzylation of benzyl chloride with potassium 3-ethoxy-2, 2-dimethyl-3-oxopropanoate, but obtained the benzyl ester without decarboxylation in high yield (Scheme 8.4). The formation of benzyl ester from the α-ester salt is possibly ascribed to the different mechanism of decarboxylation under palladium catalysis compared with the α-cyano substrate and merits further investigation.

8.2.3 Proposed Reaction Mechanism

Although the detailed mechanism of the decarboxylative benzylation is not clear at present, based on the observations above and our previous studies [20], we consider

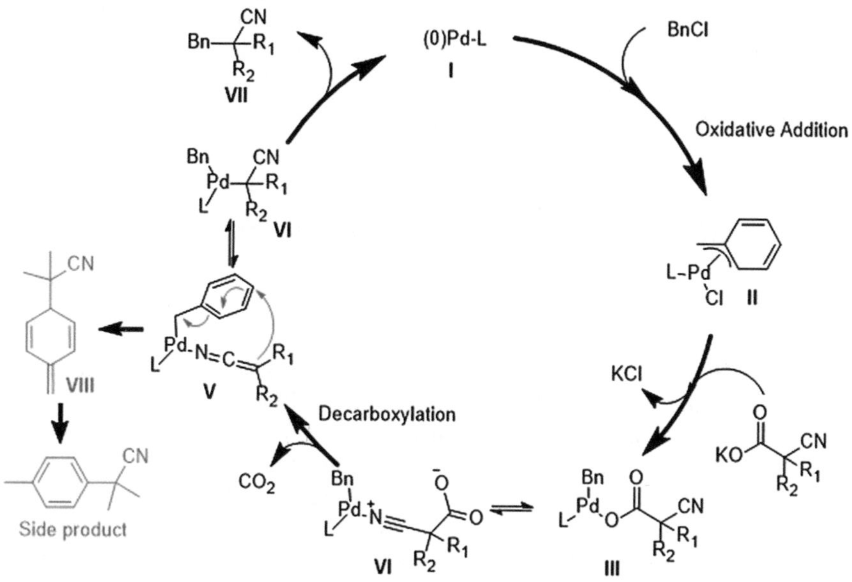

Fig. 8.2 Proposed mechanism of the decarboxylative benzylation. Reproduced from Adv. Syn. Catal. 2012, 354, 2465 by permission of John Wiley & Sons Ltd.

that the outlined mechanism may contain the following steps (Fig. 8.2). First, oxidative addition of benzyl electrophile forms a phosphine ligated benzyl-Pd complex II. The carboxylate replaces the halide through anion exchange to form intermediate III. Intermediate III isomerizes to intermediate IV, in which the palladium coordinates to the cyano nitrogen. The carboxylate decarboxylates on IV to form a nitrile-nitrogen coordinated intermediate V. V isomerizes to C-coordinated style VI followed by reductive elimination to give the desired product. Notably, in intermediate V, the α-carbon of the nitrile could attack the *para* site of the benzyl group to produce a dearomatized product VIII followed by rearomatization to generate the observed by-product. The formation of the arylation by-product may give evidence that palladium interacts with the nitrile-nitrogen in the catalytic process.

8.3 Conclusion

In summary, an efficient and practical palladium-catalyzed decarboxylative benzylation of α-cyano aliphatic carboxylate salts with benzyl electrophiles has been developed. This reaction avoids the use of sensitive reagents, proceeds under relatively mild conditions, and is compatible to various functional groups. A diverse range of quaternary, tertiary, and secondary β-aryl nitriles can be conveniently prepared using this reaction. Many of these nitriles are difficult to synthesize via traditional base-mediated nucleophilic substitutions. Further efforts to elucidate the detailed mechanism and achieve an asymmetric decarboxylative benzylation of quaternary substrates will be the future goal of our research.

8.4   Experimental Section and Compound Data

8.4.1   General Information

All the reactions were carried out in oven-dried Schlenk tubes under Argon atmosphere (purity ≥99.999%). The toluene and mesitylene solvents were bought from TCI and used without further purification. Benzyl electrophiles were bought from Alfa Aesar and TCI or synthesized according to literature procedures. Allyl palladium chloride dimer was purchased from Sigma-Aldrich and Sinocompound. All phosphine ligands were bought from Sinocompound, Sigma-Aldrich, Strem, or Alfa Aesar and used as is. All the other reagents and solvents mentioned in this text were bought from Sinopharm Chemical Reagent Co. Ltd or Alfa Aesar and purified when necessary.

^{1}H-NMR, ^{13}C-NMR spectra were recorded on a Bruker Avance 400 spectrometer at ambient temperature in CDCl$_3$ unless otherwise noted. Data for ^{1}H-NMR are reported as follows: chemical shift (δ ppm), multiplicity, integration, and coupling constant (Hz). Data for ^{13}C-NMR are reported in terms of chemical shift (δ ppm), multiplicity, and coupling constant (Hz). Gas chromatographic (GC) analysis was acquired on a Shimadzu GC-2014 Series GC System equipped with a flame-ionization detector. GC-MS analysis was performed on Thermo Scientific AS 3000 Series GC-MS System. HRMS analysis was performed on Finnigan LCQ advantage Max Series MS System. Organic solutions were concentrated under reduced pressure on a Buchi rotary evaporator. Flash column chromatographic purification of products was accomplished using forced-flow chromatography on Silica Gel (200–300 mesh).

8.4.2   Experimental Procedure

Optimization of the Reaction Conditions and Survey of the Benzyl Electrophiles

A 10 mL oven-dried Schlenk tube was charged with [PdCl(allyl)]$_2$ (2% mol), ligand (6% mol, see SI Table 8.5), α-cyano aliphatic carboxylate salts (see Table 8.5, 0.375 mmol), and the corresponding benzyl electrophile (0.25 mmol, if solid, See Table 8.5). The tube was evacuated and filled with argon (this procedure was repeated for three times). Then benzyl electrophile (0.25 mmol, if liquid, see

Table 8.5 Pd-catalyzed decarboxylative benzylation of benzyl chloride with potassium 2-cyano-2-methylpropanoate[a]

Entry	M	T (°C)	Ligand	Solvent	Yield	
					1	2
1	K	140	Xant-Phos	Mesitylene	8	53
2	K	140	BINAP	Mesitylene	54	9
3	K	140	DPPF	Mesitylene	80	Trace
4	K	140	DPE-Phos	Mesitylene	56	Trace
5	K	140	DPPPent	Mesitylene	71	–
6	K	140	CataCXium A	Mesitylene	18	–
7	K	140	Ph-Dave-Phos	Mesitylene	78	–
8	K	140	t-BuX-Phos	Mesitylene	32	–
9	K	140	Ru-Phos	Mesitylene	95	–
10	K	140	X-Phos	Mesitylene	94	–
11	K	140	S-Phos	Mesitylene	98	–
12	K	140	IPr·HCl	Mesitylene	74	–
13	Na	140	S-Phos	Mesitylene	94	–
14[b]	Li	140	S-Phos	Mesitylene	Trace	–
15	K	140	S-Phos	DMA	51	–
16	K	140	S-Phos	Diglyme	95	–
17	K	140	S-Phos	NMP	57	–
18	K	140	S-Phos	DMSO	20	6
19	K	110	S-Phos	Mesitylene	95	–
20	K	100	S-Phos	Mesitylene	95	–
21	K	110	S-Phos	Toluene'	95	–

[a]GC yields using biphenyl as the internal standard (average of two runs). [b]recovery of the benzyl chloride in 90% yield

Table 8.5) and solvent (0.50 mL, see Table 8.5) were added with a syringe under a counter flow of argon. The tube was sealed with a screw cap, stirred at room temperature for 1 min, and connected to the Schlenk line, which was filled with argon, stirred in a preheated oil bath at mentioned temperature (see Table 8.5) for 5 h. Upon completion of the reaction, the mixture was cooled to room temperature and diluted with ethyl acetate. The yields were determined by gas chromatography using biphenyl as internal standard.

The structures of the ligands mentioned in Table 8.5 are shown in Fig. 8.3.

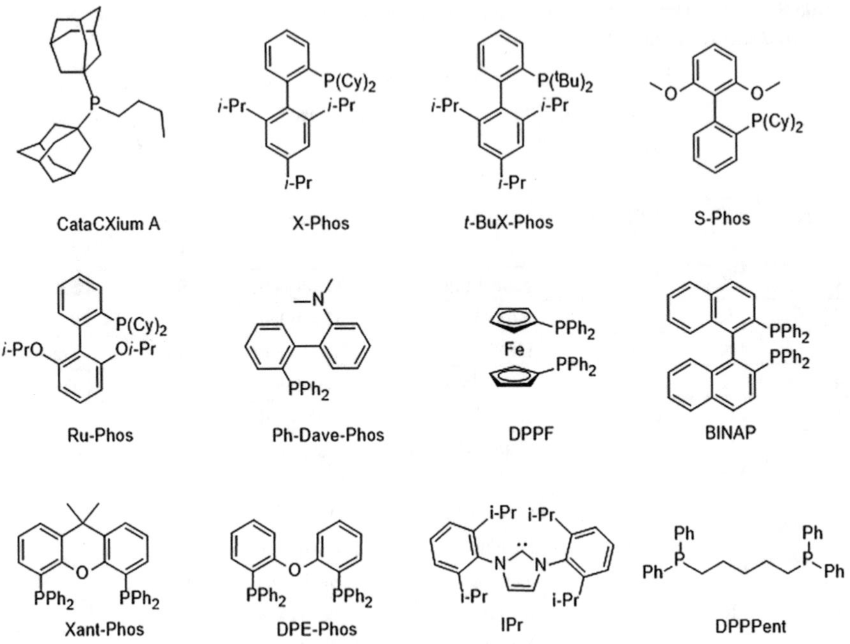

| CataCXium A | X-Phos | t-BuX-Phos | S-Phos |

| Ru-Phos | Ph-Dave-Phos | DPPF | BINAP |

| Xant-Phos | DPE-Phos | IPr | DPPPent |

Fig. 8.3 Structures of the ligands mentioned in Table 8.5

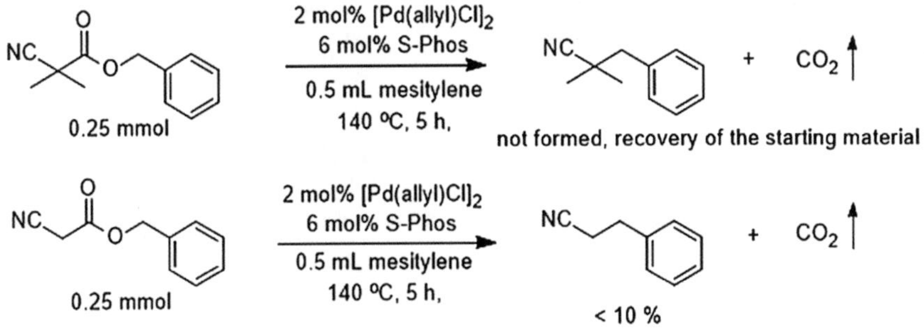

Scheme 8.5 Tentative intramolecular decarboxylative benzylation of benzyl esters

Control Experiments and Some Unexpected Results

Control experiment to test the intramolecular decarboxylative coupling of benzyl 2-cyanoacetate was unsuccessful under the optimized conditions (Scheme 8.5). This reaction absolutely requires a palladium catalyst (Scheme 8.6).

When 4-fluorobenzyl chloride was used as the substrate, the expected decarboxylative benzylation product was detected in only trace amount, while 2-methyl-2-(p-tolyl)propanenitrile, a product formed through C–H activation and

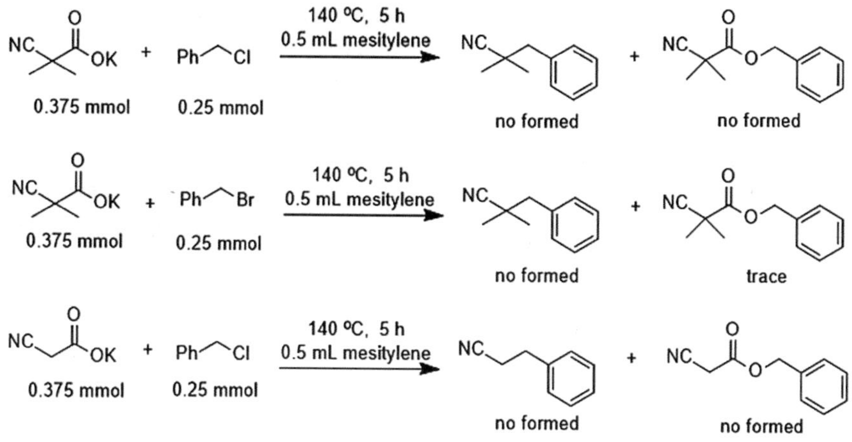

Scheme 8.6 Control experiments in the absence of palladium catalyst

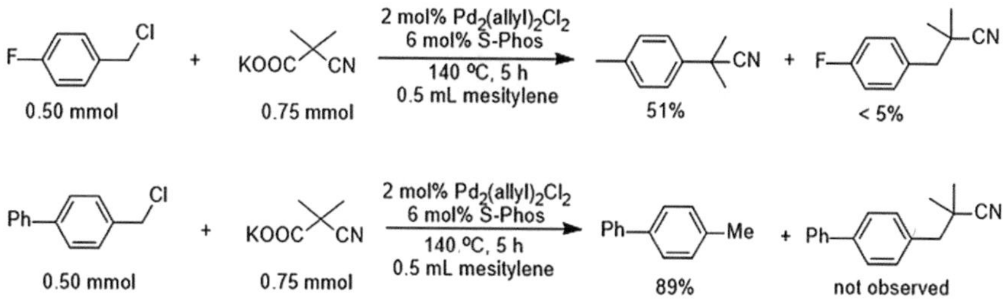

Scheme 8.7 Some unexpected results

dechlorination was detected in 51%. When 4-phenyl benzyl chloride was used as the substrate, only dehalogenation took place to generate 4-methyl-biphenyl (Scheme 8.7).

8.4.3   Characterization of the Products

Compound name 2,2-dimethyl-3-(o-tolyl)propanenitrile

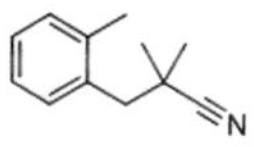

81 mg (94%) of a yellow oil was obtained. This compound is known. Reference: U. S. Pat. Appl. Publ. (2010), US 20100137178 A1 20100603. ^{1}H NMR (400 MHz, CDCl$_3$) δ 7.29–7.25 (m, 1H), 7.19–7.14 (m, 3H), 2.85 (s, 2H), 2.36 (s, 3H), 1.37 (s, 6H). ^{13}C NMR (101 MHz, CDCl$_3$) δ 136.96, 134.15, 130.99, 130.74, 127.30, 125.83, 124.97, 42.35, 33.83, 26.79, 20.19.

Compound name 2,2-dimethyl-3-(3-nitrophenyl)propanenitrile

34 mg (33%) of a yellow solid was obtained. This compound is known. Reference: Journal of Organic Chemistry (2002), 67(26), 9428–9438. ^{1}H NMR (400 MHz, CDCl$_3$) δ 8.18 (d, J = 8.2 Hz, 1H), 8.12 (s, 1H), 7.69 (d, J = 7.6 Hz, 1H), 7.56 (t, J = 7.9 Hz, 1H), 2.94 (s, 2H), 1.40 (s, 6H). ^{13}C NMR (101 MHz, CDCl$_3$) δ 148.22, 137.69, 136.28, 129.54, 124.99, 123.92, 122.58, 46.13, 33.49, 26.47.

Compound name 3-(3-fluorophenyl)-2,2-dimethylpropanenitrile

74 mg (83%) of a yellow oil was obtained. This compound is new. ^{1}H NMR (400 MHz, CDCl$_3$) δ 7.34–7.27 (m, 1H), 7.07 (d, J = 7.6 Hz, 1H), 7.03–6.96 (m, 2H), 2.80 (s, 2H), 1.36 (s, 6H). ^{13}C NMR (101 MHz, CDCl$_3$) δ 162.66 (d, J_F = 246.2 Hz), 138.09 (d, J_F = 7.3 Hz), 129.90 (d, J_F = 8.3 Hz), 125.91 (d, J_F = 2.9 Hz), 124.44, 117.09 (d, J_F = 21.3 Hz), 114.35 (d, J_F = 20.9 Hz), 46.30 (d, J_F = 1.7 Hz), 33.41, 26.52. HRMS calcd for C$_{11}$H$_{12}$FN (M+) 177.0948; found: 177.0939.

Compound name 2,2-dimethyl-3-(pyridin-3-yl)propanenitrile

24 mg (30%) of a yellow oil was obtained. This compound is new. ^{1}H NMR (400 MHz, CDCl$_3$) δ 8.58–8.55 (m, 1H), 8.49 (s, 1H), 7.70 (d, J = 7.8 Hz, 1H), 7.30 (dd, J = 7.8, 4.8 Hz, 1H), 2.82 (s, 2H), 1.38 (s, 6H). ^{13}C NMR (101 MHz, CDCl$_3$) δ 151.10, 148.93, 137.49, 131.28, 124.11, 123.44, 43.82, 33.41, 26.42. HRMS calcd for C$_{10}$H$_{12}$N$_2$ (M+) 160.1000; found: 160.0996.

Compound name 2,2-dimethyl-3-(thiophen-3-yl)propanenitrile

72 mg (87%) of a yellow oil was obtained. This compound is new. ^{1}H NMR (400 MHz, CDCl$_3$) δ 7.28 (dd, J = 4.9, 3.0 Hz, 1H), 7.15–7.12 (m, 1H), 7.05 (dd, J = 4.9, 1.3 Hz, 1H), 2.85 (s, 2H), 1.33 (s, 6H). ^{13}C NMR (101 MHz, CDCl$_3$) δ 136.02, 129.10, 125.55, 124.93, 123.75, 41.05, 33.28, 26.38. HRMS calcd for C$_9$H$_{11}$NS (M+) 165.0607; found: 165.0604.

Compound name 2,2-dimethyl-3-(3-(trifluoromethyl)phenyl)propanenitrile

83 mg (73%) of a yellow oil was obtained. This compound is new. ^{1}H NMR (400 MHz, CDCl$_3$) δ 7.57 (d, J = 7.3 Hz, 1H), 7.52–7.45 (m, 3H), 2.87 (s, 2H), 1.36 (s, 6H). ^{13}C NMR (101 MHz, CDCl$_3$) δ 136.65, 133.56 (d, J_F = 1.1 Hz), 130.80 (q, J_F = 32.4 Hz), 129.01, 126.90 (q, J_F = 3.8 Hz), 124.33 (q, J_F = 3.8 Hz), 124.06 (q, J_F = 272.3 Hz), 124.20, 46.38, 33.47, 26.46. HRMS calcd for C$_{12}$H$_{12}$F$_3$N (M+) 227.0916; found: 227.0911.

Compound name (E)-butyl 3-(3-(2-cyano-2-methylpropyl)phenyl)acrylate

127 mg (89%) of a yellow oil was obtained. This compound is new. ^{1}H NMR (400 MHz, CDCl$_3$) δ 7.68 (d, J = 16.0 Hz, 1H), 7.47 (d, J = 7.5 Hz, 1H), 7.41 (s, 1H), 7.36 (t, J = 7.6 Hz, 1H), 7.31 (d, J = 7.6 Hz, 1H), 6.45 (d, J = 16.0 Hz, 1H), 4.21 (t, J = 6.7 Hz, 2H), 2.82 (s, 2H), 1.73–1.65 (m, 2H), 1.49 – 1.39 (m, 2H), 1.36 (s, 6H), 0.96 (t, J = 7.4 Hz, 3H). ^{13}C NMR (101 MHz, CDCl$_3$) δ 166.94, 144.18, 136.47, 134.67, 131.98, 129.83, 128.98, 126.98, 124.44, 118.75, 64.43, 46.45, 33.45, 30.78, 26.51, 19.19, 13.74. HRMS calcd for C$_{18}$H$_{23}$NO$_2$ (M+) 285.1723; found: 285.1716.

Compound name 2,2-dimethyl-3-(3-((5-(4,4,5,5-tetramethyl-1,3,2-dioxaborolan-2-yl)pentyl)oxy)phenyl)propanenitrile

165 mg (89%) of a yellow oil was obtained. This compound is new. ^{1}H NMR (400 MHz, CDCl$_3$) δ 7.22 (t, J = 7.7 Hz, 1H), 6.86–6.78 (m, 3H), 3.94 (t, J = 6.6 Hz, 2H), 2.77 (s, 2H), 1.78 (p, J = 6.8 Hz, 2H), 1.52–1.43 (m, 4H), 1.34 (s, 6H), 1.24 (s, 12H), 0.81 (t, J = 7.3 Hz, 2H). ^{13}C NMR (101 MHz, CDCl$_3$) δ 159.11, 137.04, 129.25, 124.85, 122.33, 116.47, 113.35, 82.90, 67.89, 46.70, 33.38, 29.08, 28.73, 26.58, 24.83, 23.80, 11.12. HRMS calcd for C$_{22}$H$_{34}$BNO$_3$ (M+) 371.2626; found: 371.2616.

Compound name 3-(3-(3-chloropropoxy)phenyl)-2,2-dimethylpropanenitrile

107 mg (85%) of a yellow oil was obtained. This compound is new. ^{1}H NMR (400 MHz, CDCl$_3$) δ 7.26–7.21 (m, 1H), 6.87–6.81 (m, 3H), 4.12 (t, J = 5.8 Hz, 2H), 3.74 (t, J = 6.4 Hz, 2H), 2.78 (s, 2H), 2.26 – 2.20 (m, 2H), 1.35 (s, 6H). ^{13}C

NMR (101 MHz, CDCl$_3$) δ 158.68, 137.26, 129.37, 124.80, 122.83, 116.43, 113.40, 64.27, 46.67, 41.52, 33.44, 32.29, 26.59. HRMS calcd for C$_{14}$H$_{18}$ClNO (M +) 251.1071; found: 251.1074.

Compound name 4-benzyltetrahydro-2H-pyran-4-carbonitrile

78 mg (78%) of a yellow oil was obtained. This compound is new. ^{1}H NMR (400 MHz, CDCl$_3$) δ 7.39 – 7.26 (m, 5H), 3.94 (dd, J = 12.0, 4.0 Hz, 2H), 3.67 (td, J = 12.2, 2.2 Hz, 2H), 2.87 (s, 2H), 1.78 (dd, J = 13.6, 1.2 Hz, 2H), 1.68 (td, J = 12.8, 4.4 Hz, 2H). ^{13}C NMR (101 MHz, CDCl$_3$) δ 134.19, 130.29, 128.48, 127.54, 122.16, 64.63, 46.20, 38.13, 35.30. HRMS calcd for C$_{13}$H$_{15}$NO (M+) 201.1154; found: 201.1150.

Compound name (E)-butyl 3-(3-(2-cyano-2-methyl-4-oxo-4-phenylbutyl)phenyl) acrylate

142 mg (73%) of a yellow oil was obtained. This compound is new. ^{1}H NMR (400 MHz, CDCl$_3$) δ 7.90 (d, J = 7.2 Hz, 2H), 7.63 (d, J = 16.0 Hz, 1H), 7.59 (tt, J = 7.6, 1.2 Hz, 1H), 7.49–7.40 (m, 4H), 7.38 – 7.31 (m, 2H), 6.38 (d, J = 16.0 Hz, 1H), 4.19 (t, J = 6.7 Hz, 2H), 3.30–3.14 (m, 3H), 3.04 (d, J = 13.5 Hz, 1H), 1.74 – 1.61 (m, 2H), 1.54 (s, 3H), 1.49 – 1.35 (m, 2H), 0.96 (m, J = 7.4 Hz, 3H). ^{13}C NMR (101 MHz, CDCl$_3$) δ 195.44, 166.89, 144.01, 136.50, 135.94, 134.79, 133.74, 132.18, 129.97, 129.11, 128.82, 127.92, 127.20, 123.35, 118.82, 64.43, 44.93, 43.84, 35.05, 30.76, 24.45, 19.18, 13.74. HRMS calcd for C$_{25}$H$_{27}$NO$_3$ (M+) 389.1985; found: 389.1988.

Compound name ethyl 2-(3-(2,7-dicyano-2-methylheptyl)phenoxy)acetate

113 mg (66%) of a yellow oil was obtained. This compound is new. ^{1}H NMR (400 MHz, CDCl$_3$) δ 7.28 – 7.23 (m, 1H), 6.91 (d, J = 7.7 Hz, 1H), 6.87 – 6.80 (m, 2H), 4.62 (s, 2H), 4.27 (q, J = 7.2 Hz, 2H), 2.88 (d, J = 13.5 Hz, 1H), 2.71 (d, J = 13.5 Hz, 1H), 2.36 (t, J = 7.0 Hz, 2H), 1.78 – 1.39 (m, 8H), 1.30 (t, J = 7.0 Hz, 3H), 1.27 (s, 3H). ^{13}C NMR (101 MHz, CDCl$_3$) δ 168.82, 157.84, 137.01, 129.52, 123.79, 123.70, 119.51, 116.85, 113.49, 65.43, 61.38, 45.41, 39.04, 37.73, 28.61, 25.17, 24.38, 23.88, 17.10, 14.17. HRMS calcd for C$_{20}$H$_{26}$N$_2$O$_3$ (M +) 342.1938; found: 342.1930

References

1. (a) Goossen, L. J., Rodriguez, N., & Goossen, K. (2008). *Angewandte Chemie International Edition, 47*, 3100–3120. (b) Weaver, J. D., Recio, A., III, Grenning, A. J., & Tunge, J. A. (2011). *Chemical Reviews, 111*, 1846–1913. (c) Shang, R., & Liu, L. (2011). *Science China Chemistry, 54*, 1670–1687.
2. (a) Myers, A. G., Tanaka, D., & Mannion, M. R. (2002). *Journal of the American Chemical Society, 124*, 11250–11251. (b) Tanaka, D., Romeril, S. P., & Myers, A. G. (2005). *Journal of the American Chemical Society, 127*, 10323–10333.
3. (a) Goossen, L. J., Deng, G., & Levy, L. M. (2006). *Science, 313*, 662–664. (b) Goossen, L. J., Rodriguez, N., Melzer, B., Linder, C., Deng, G., & Levy, L. M. (2007). *Journal of the American Chemical Society, 129*, 4824–4833. (c) Goossen, L. J., Rodriguez, N., & Linder, C. (2008). *Journal of the American Chemical Society, 130*, 15248–15249.
4. Forgione, P., Brochu, M. C., St-Onge, M., Thesen, K. H., Bailey, M. D., & Bilodeau, F. (2006). *Journal of the American Chemical Society, 128*, 11350–11351.
5. (a) Wang, Z., Ding, Q., He X., & Wu (2009). *Organic & Biomolecular Chemistry, 7*, 863–865. (b) Yamashita, M., Hirano, K., Satoh, T., & Miura, M. (2010). *Organic Letters, 12*, 592–595.
6. Moon, J., Jeong, M., Nam, H., Ju, J., Moon, J. H., Jung, H. M., & Lee, S. (2008). *Organic Letters, 10*, 945–948.
7. (a) Shang, R., Yang, Z. W., Wang, Y., Zhang, S.-L., & Liu, L. (2010). *Journal of the American Chemical Society, 132*, 14391–14393. (b) Shang, R., Ji, D. S., Chu, L., Fu, Y., & Liu, L. (2011). *Angewandte Chemie International Edition, 50*, 4470–4474. (c) Yeung, P. Y., Chung, K. H., & Kwong, F. Y. (2011). *Organic Letters, 13*, 2912–2915. (d) Shang, R., Huang, Z., Chu, L., Fu, Y., & Liu, L. (2011). *Organic Letters, 13*, 4240–4243. (e) Chou, C-M., Chatterjee, I., & Studer, A. (2011). *Angewandte Chemie International Edition, 50*, 8614–8617.
8. Rayabarapu, D. K., & Tunge, J. A. (2005). *Journal of the American Chemical Society, 127*, 13510–13511.
9. Torregrosa, R. R. P., Ariyarathna, Y., Chattopadhyay, K., & Tunge, J. A. (2010). *Journal of the American Chemical Society, 132*, 9280–9282.
10. Milstein, D., & Stille, J. K. (1979). *Journal of the American Chemical Society, 101*, 4992–4998.
11. (a) Taber, D. F., & Kong, S. (1997). *The Journal of Organic Chemistry, 62*, 8575–8576. (b) Fleming, F. F., & Shook, B. C. (2002). *Tetrahedron, 58*, 1–23.
12. Diez-Barra, F., De la Hoz, A., Moreno, A., & Sanchez-Verdu, P. (1989). *Synthesis*, 391–393.
13. Fleming, F. F., Yao, L., Ravikumar, P. C., Funk, L., & Shook, B. C. (2010). *Journal of Medicinal Chemistry, 53*, 7902–7917.
14. Wendy, H. F., & Chruma, J. J. (2010). *Organic Letters, 12*, 316–319.
15. It is very intereting that when 4-phenyl benzyl chloride was used as the substrate, only dehalogenation took place to generate 4-methyl-biphenyl.
16. Jana, R., Pathak, T. P., & Sigman, M. S. (2011). *Chemical Reviews, 111*, 1417–1492.
17. Moradi, W. A., & Buchwald, S. L. (2001). *Journal of the American Chemical Society, 123*, 7996–8002.
18. Control experiments shown that no decarboxylative benzylated product was observed in the absence of the palladium catalyst.
19. Bao, M., Nakamura, H., & Yamamoto, Y. (2001). *Journal of the American Chemical Society, 123*, 759–760.
20. Jiang, Y., Fu, Y., & Liu, L. (2012). *Science China Chemistry, 55*, 2057–2062.

Part II
New Carbon–Carbon Coupling Reactions Based on Iron-Catalyzed C–H Activation

Chapter 9
Recent Developments of Iron-Catalyzed Directed C–H Activation/C–C Bond Formation Reactions

Abstract Driven by the interest in chemical utilization of ubiquitous metals that are abundant and non-toxic, iron catalysis has become a rapidly growing area of research, and iron-catalyzed C–H activation is actively explored in recent years. In this chapter, I summarize the developments of iron-catalyzed C–H activation that emerged recently.

9.1 Introduction

It is an ideal way to construct a C–C bond by selective C–H bond activation. First, compared with the traditional cross-coupling concept, direct activation of C–H bond to build the C–C bond avoids the prefunctionalization, which meets the need of economic and atomic economy. Second, because C–H bonds are widely existed in the organic molecules, the highly selective C–H bond activation methods leads to the highly selective C–C bond formation, which provides direct and highly regioselective ways to build C–C bonds, thus will significantly improve the ability to access complex molecules in organic chemistry. In the past two decades, the noble metal (palladium, ruthenium, rhodium, and iridium) catalyzed C–H bond activation reactions have received intensive attentions by chemist worldwide. Since early reports reported that many noble metals have the ability to cleave the C–H bonds, the majority of the C–H bond activation and C–C bond formation research are in the frame of using catalyst based on noble metals [1]. However, most of these precious metals are rather toxic, rare, and are nonrenewable resources. Standing on the time scale of centuries, the consumption of precious metals resources will inevitably lead to the scarce of the precious metals resources, followed by rising price. The global distribution of precious metal resources is uneven, the price is also affected by political and economic factors, as Japanese government approved the program of "Elements Strategy Initiative" [2]. With these problems in mind, iron is nontoxic and the most abundant transition-metal elements on earth to conform the criteria for sustainable catalysis. However, research about iron-catalysis is still rare compared with precious metal catalysis [3]. Development of novel iron-based

© Springer Nature Singapore Pte Ltd. 2017
R. Shang, *New Carbon–Carbon Coupling Reactions Based on Decarboxylation and Iron-Catalyzed C–H Activation*, Springer Theses,
DOI 10.1007/978-981-10-3193-9_9

catalyst to realize C–H conversion reaction in place of the noble metal catalysis, exploring the unknown catalytic activity of iron and achieving unprecedent new transformations are the challenges but also opportunities for chemists. In the first chapter of the second part of the thesis, we summarize the current development of iron-catalyzed C–H bond activation and C–C bond formation reactions. The discussion is classified by different types of C–C bond formed after C–H activations. The problems need to be solved and opportunities in this area are also discussed.

9.2   Iron-Catalyzed Direct Arylation of the C–H Bonds with Aryl Halides as the Arylation Reagents

In 2010, Charette reported the iron-catalyzed arylation of aryl C–H bond with aryl iodides and aryl bromides [4] (Scheme 9.1). The reaction does not require organometallic reagents and the assistance of directing group. Aromatic C–H bonds can be directly arylated by the aryl iodides in high yield generating biaryl compounds. In this report, $Fe(OAc)_2$ was selected as the catalyst and 4,7-bathophenanthroline as optimal ligand for this reaction.

Aryl bromides also can be used in this reaction, while the yield is moderate. Arene substrates in this reaction are used as the solvent, thus severely limited the scope of arenes applicable. The regioselectivity of arylation for substituted arenes are moderate. For example, when trimethylsilyl substituted benzene was used as substrate, the arylation product was observed in low yield as a mixture of regioisomers ($o:m:p = 1.0:1.4:2.0$). Considering the separation issue, the reaction is only synthetically useful for arylation of symmetric arenes that possess identical C–H bond (Scheme 9.2).

Mechanistic studies by the same authors disclosed that C–H bond cleavage is not the rate-determining step in this reaction, and the radical scavenger, such as TEMPO, suppresses this reaction. Based on their results of mechanistic studies, the authors proposed a possible mechanism for this reaction. The authors indicated that this reaction proceeds through iron-catalyzed hemolytic radical substitution (Scheme 9.3). First, Fe(II) catalyst (**A**) interact with the aryl halides through a single electron transfer (SET) to form Fe(III) and an aryl radical intermediates (**B**), radical intermediates (**B**) adds to arene to give a new radical that may be stabilized by Fe(III) center (**C**). Potassium tert-butoxide acts as a base to deprotonate intermediate (**C**), and oxidation of the carbon radical by Fe(III) produces biaryl

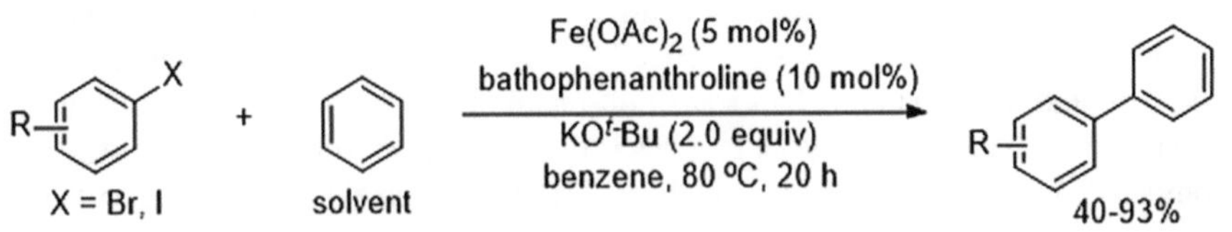

Scheme 9.1 Iron-catalyzed direct arylation of benzene with aryl halides

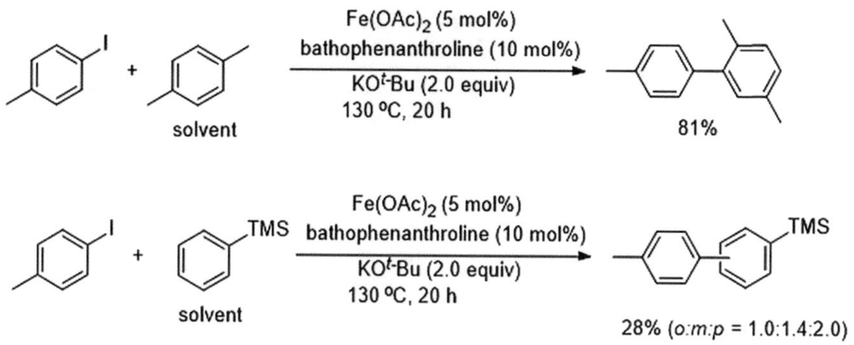

Scheme 9.2 Iron-catalyzed direct arylation of arenes with 4-tolyl iodide

Scheme 9.3 Possible mechanism of iron-catalyzed direct arylation of arenes with aryl iodide

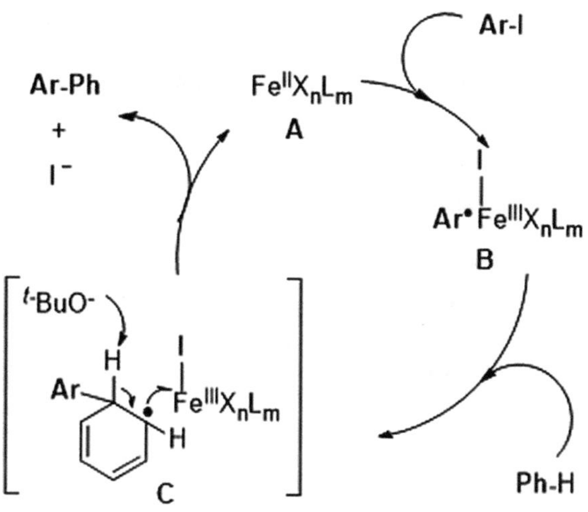

compounds and regenerates iron catalyst. According to this proposed mechanism, this process involves iron intermediates possessing radical property or simply an outsphere carbon radical is involved. The mechanism of this iron-catalyzed arylation is distinct from Pd or Ru-catalyzed similar processes.

Lei et al. subsequently reported a similar iron-catalyzed C–H arylation reaction [5]. In Lei's work, $FeCl_3$ was chosen as optimal catalyst that is distinct from Charette's system using Fe(II). The authors also found that N,N'-dimethylethylenediamine (DMEDA) acts as a ligand to promote this reaction (Scheme 9.4). In Lei's report, boht aryl iodides and aryl bromides can be used to obtain good yields, but the use of a stronger and moisture sensitive base, hexamethyldisilazide lithium amide (LiHMDS), is necessary (Scheme 9.5). It is possible that the base reduces Fe (III) to generate Fe(II) as catalyst through single electron transfer in the reaction

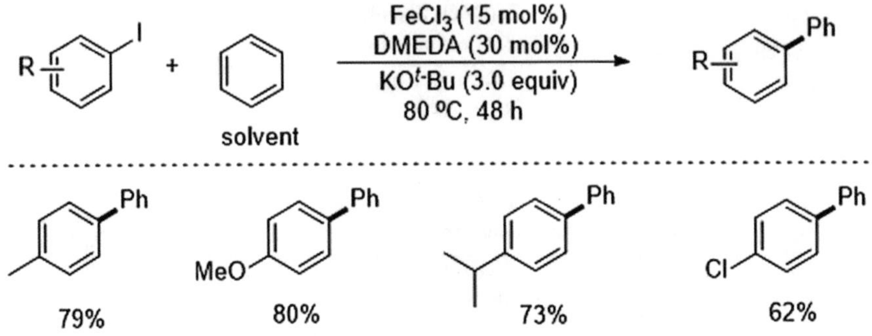

Scheme 9.4 FeCl₃/DMEDA-catalyzed direct arylation of benzene with aryl iodides

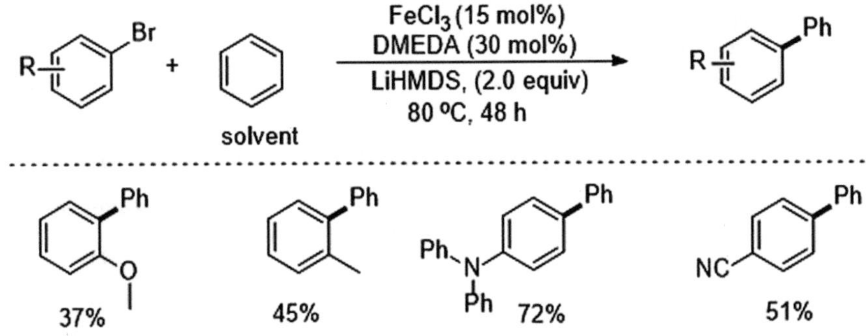

Scheme 9.5 FeCl₃/DMEDA-catalyzed direct arylation of benzene with aryl bromides

system. Aryl chlorides were proved to be reactive, while the yields are much lower compared with other halides (Br and I). Aryl halides containing ortho substitution are amenable in Lei's catalytic system. The applicability of the reaction was also limited by the necessity of using a large excess of arene. The reaction delivers a mixture of regioisomers with low regioselectivity when arylation of substituted arenes are required.

It is worth mentioning that although the C–H arylation reaction is reported to proceed in the presence of and iron catalyst, Shi [6], Lei [7], and Hayashi [8] later pointed out that the iron catalyst is not necessary for this process to take place. Subsequent studies revealed this type of reaction proceeded well when only TMEDA or 1,10-Phenanthroline was used as catalyst in the presence of a base. Thus, it is more likely this type of reaction is actually a radical type homolytic aromatic substitution rather than proceeds through organometallic intermediate possessing Fe–C bonds.

9.3 Iron-Catalyzed Direct Activation of C–H Bond Using Organometallic Reagents

Except for the iron-catalyzed arylation reaction using aryl halides, iron-catalyzed direct C–H bond arylation reaction using organometallic reagents have also been developed. In 2008, Nakamura and co-workers reported a novel iron-catalyzed direct C–H bond arylation reaction using arylzinc reagent (Scheme 9.6). Using organozinc reagent prepared from Grignard reagent, arylation of α-benzoquinoline can be achieved with a high selectivity at the C10 position in high yield [9]. Obviously, an oxidant is necessary for this reaction. In this report, the authors disclosed that 1,2-dichloroisobutane (DCIB) is efficient to promote the turnover of iron catalyst. 1,2-dihaloalkane was previously reported to be an effective oxidant for iron-catalyzed oxidative coupling reactions. The authors screened various types of ligands and pointed out dinitrogen ligands, such as 1,10-phenanthroline, are most effective for this reaction. Compared with ruthenium or palladium-catalyzed reactions, this reaction can occur at much lower temperature. The optimal result demonstrated that the arylation product can be obtained in 99% yield at 0 °C after reaction for 16 h, indicating the high reactivity of organoiron catalyst to cleave C–H bond. The drawback of the reaction is that excess amount of organometallic reagents is used, which lowers the functional group compatibility. The author revealed that only one phenyl group in the diaryl Zinc reagent can participate in the reaction, since the reaction was entirely failed when PhZnBr was used. Among the three equivalents of diarylzinc reagent, one equivalent of diphenyl zinc acts as base to deprotonate C–H bond and the other two equivalents act as coupling partner to form C-aryl bond. The excess amount of diarylzinc reagent is necessary due to the concurrence of undesired homocoupling of organometallic reagent. The reaction opens a new window for iron-catalysis and reveals the future opportunity to use sustainable iron catalyst in step-efficient C–H transformations.

For the scope of substrates (Scheme 9.7), Nakamura et al. demonstrated that this reaction is applicable for a variety of 2-aryl-pyridines containing both electron-withdrawing and electron-donating substituents. The ester groups are compatible in the reaction. For the 2-aryl-pyridine with a steric substitution at the 8-position, the reaction delivers mono- and di-arylated products as a mixture. The reaction is very sensitive to steric hindrance, as for 2-(3-tolyl)pyridine, only mono-arylated product was obtained. Sterically hindered aryl zinc reagent such as o-tolylzinc reagent is unreactive, revealing a organoiron intermediate that is

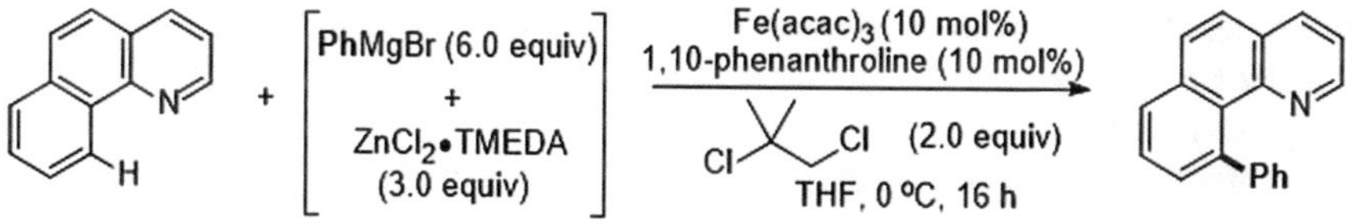

Scheme 9.6 Iron-catalyzed direct arylation of α-benzoquinoline with phenylzinc reagent

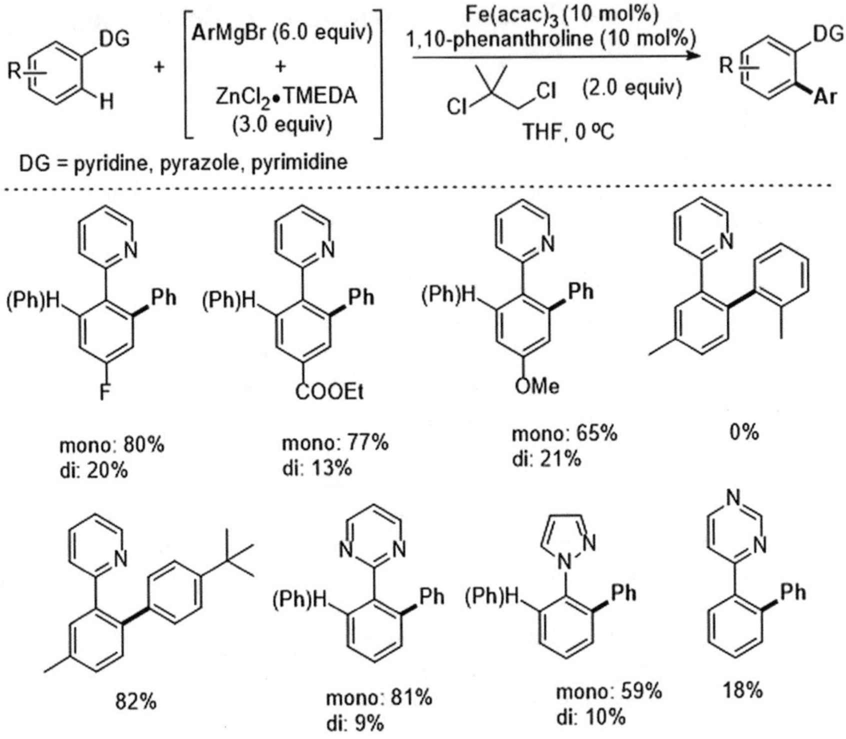

Scheme 9.7 Iron-catalyzed nitrogen directed arylation using arylzinc reagent

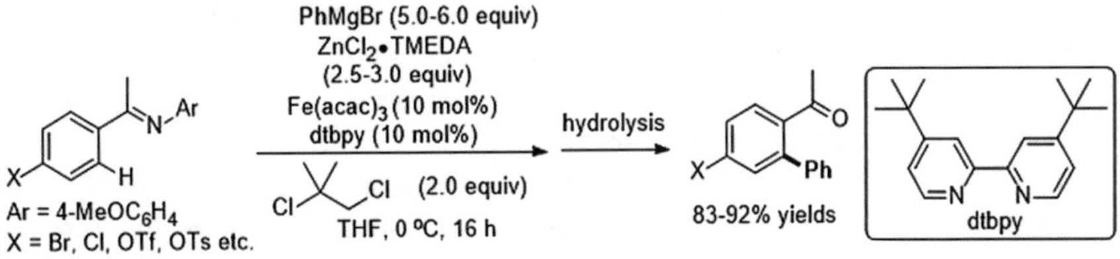

Scheme 9.8 Iron-catalyzed direct arylation of (pseudo) halogen-bearing ketimine with phenylzinc reagent

sensitive to steric hindrance may be involved. The authors also show that besides 2-aryl-pyridine, 2-phenyl pyrimidine and the *N*-phenyl pyrazole are also suitable substrates. This reaction is general for nitrogen atom directed C–H bond functionalization. For 4-phenylpyrimidine, the desired product can also be obtained, but the yield is low.

At present, the mechanism of this reaction (Scheme 9.7) is not clear yet, even the valence of iron in the catalytic cycle are still under debate. Further, efforts are still necessary to clarify the reaction mechanism.

In the following study, Nakamura and co-workers realized the nitrogen atom in aromatic ketimine and aldimine can also direct ortho-C–H bond arylation with aryl zinc reagents [10] using a similar catalyst system (Scheme 9.8). The author has shown the compatibility of C–H bond activation with aryl chloride, aryl bromide, aryl triflate, and aryl tosylate.

Daugulis and co-workers reported iron-catalyzed direct C–H alkylation reaction (Scheme 9.9) [11]. A strong base is used to deprotonate C–H bonds on the aromatic ring to generate organometallic reagents. The formed aryl organometallic reagent further cross couples with alkyl halide catalyzed by iron catalyst. The reaction can be understood as a combination of C–H deprotonative metalation with iron-catalyzed cross-coupling reaction [12].

Yu and co-workers from Sichuan University has reported one interesting example of iron-mediated C–H arylation using aryl boronic acid as the aryl source [13] (Scheme 9.10). In this reaction, the authors used a reaction system composed of heptahydrate iron sulfate and 1,4,7,10-tetraazacyclododecane. The reaction system also contains potassium phosphate as base and pyrazole as an additive. The disadvantage of this reaction is that arenes is used as solvent. Compared with other organometallic reagent, aryl boronic acid is a preferred aryl source due to its stability, diversity, and commercial availability. The reaction mechanism is not clear, but the reaction is likely to proceed though iron-mediated radical substitution.

Yu and co-workers also realized similar reaction later [14] (Scheme 9.11). Aryl boronic acid can couple with pyrrole to produce aryl substituted pyrrole in the presence of catalytic amount of iron oxalate dihydrate as catalyst and MCPA as

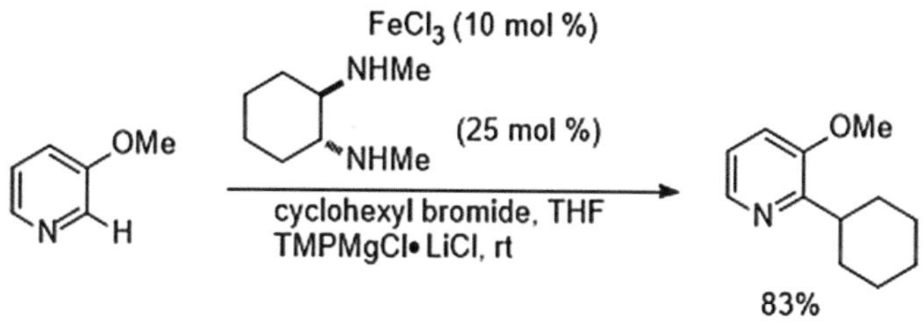

Scheme 9.9 Iron-catalyzed deprotonative alkylation of arene C–H bonds

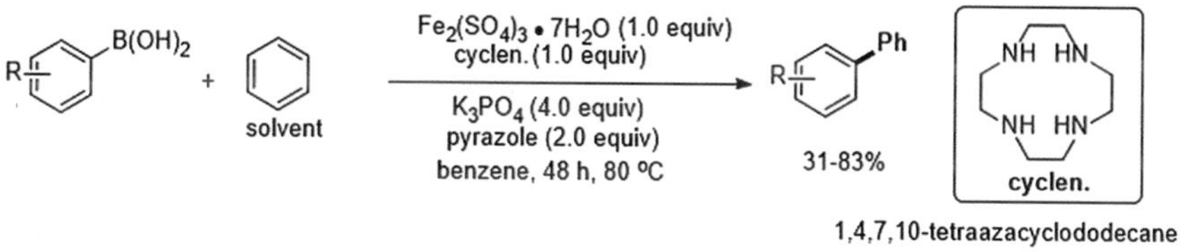

Scheme 9.10 Iron-mediated arylation of benzene with arylboronic acid

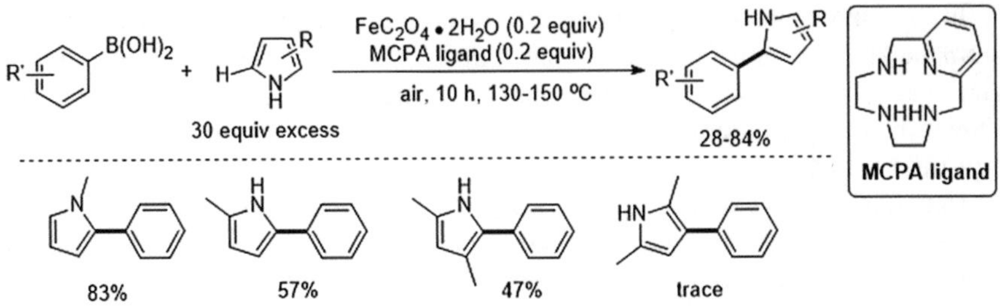

Scheme 9.11 Iron-catalyzed arylation of pyrroles with arylboronic acids

ligand. Air was used as oxidant in this reaction. The reaction can tolerate steric hindrance at ortho-position to some extent. It is noteworthy that the authors reported that the reaction only occured mainly at the 2-position of pyrrole. If the C2 position was substituted, then the reaction cannot take place at the C3-position of the pyrrole ring.

Nakamura and co-workers also reported iron-catalyzed nitrogen directed syn-arylation of alkenyl C–H bond [15] (Scheme 9.12). The homocoupling of Grignard reagent can be largely suppressed using slow addition method. The aromatic co-solvent effect was discovered to determine the syn-selectivity. When a mixed solvent of chlorobenzene and ether was used, the reaction of 2-(prop-1-en-2-yl)pyridine with phenyl magnesium bromide delivered the arylation product with high Z-selectivity (E/Z = 3:97) without stereo isomerization, when tetrahydrofuran was used as solvent, the reaction generated mainly the E- product (E/Z = 95:5). The co-solvent effect maybe explained by the coordination of aromatic solvent to low-valent iron species to inhibit an iron-catalyzed olefin Z/E isomerization. The syn- selectivity observed in this reaction rules out a mechanism of Heck process and supports a ferracycle intermediate.

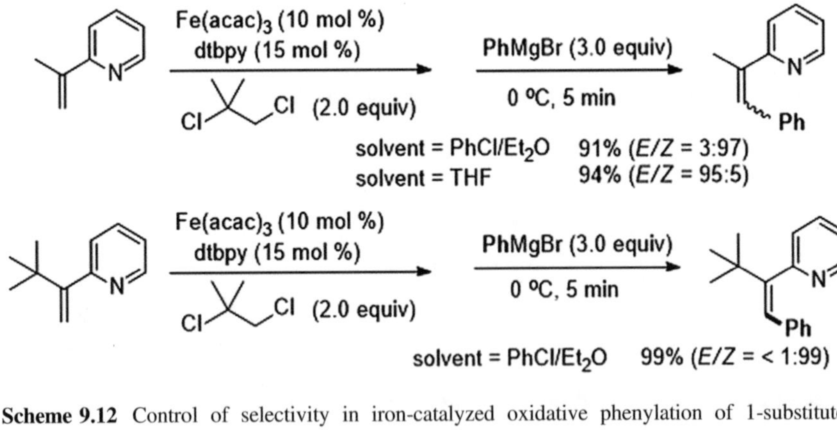

Scheme 9.12 Control of selectivity in iron-catalyzed oxidative phenylation of 1-substituted vinylpyridines

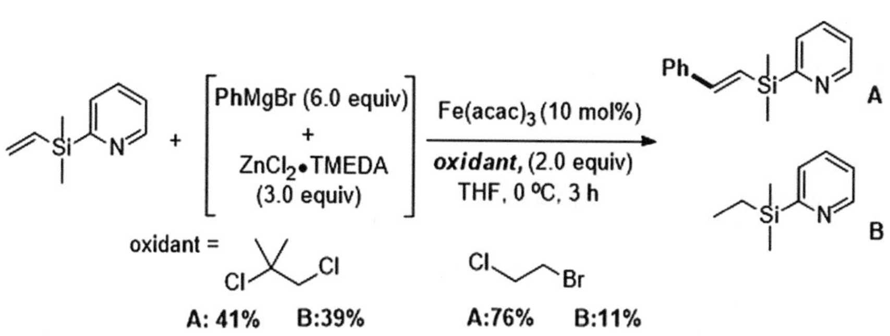

Scheme 9.13 Iron-catalyzed directed oxidative arylation of olefins with organozinc reagents

Nakamura and co-workers reported iron-catalyzed oxidative arylation of 2-(dimethyl(vinyl)silyl)pyridine using aryl zinc reagent to generate olefin product [16]. The olefin product generated in this reaction is exclusive E-isomer (Scheme 9.13). Based on this result, the authors believe that the reaction proceeds through a mechanism resembling palladium-catalyzed Mizoroki–Heck reaction, in which insertion of alkene to aryliron species followed by β-hydride elimination exists. The iron hydride species generated after β-hydride elimination is oxidized by 1,2-dihaloalkane to regenerate active iron catalyst. One interesting observation in this report is that the choice of 1,2-dihaloalkane as oxidant affects the chemoselectivity of the reaction (A, B). Compared with 1,2-dichloroisobutane, 1-bromo-2-chloroethane acts as a better oxidant to give Heck type product (A) in higher yield suppressing hydrogenation side product. This oxidant-induced chemoselectivity may be related to the rate to convert iron hydride species that cause the hydrogenation side product.

Nakamura and co-workers also reported iron-catalyzed activation of C–H bonds followed by reaction with alkynes to generate cyclization product [17] (Scheme 9.14). 2-Biaryl Grignard reagents react with alkynes in the presence of a

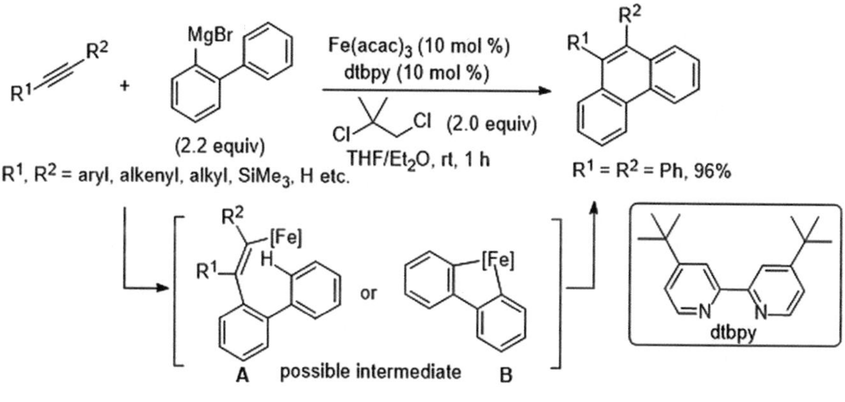

Scheme 9.14 Iron-catalyzed [4 + 2] benzannulation

catalytic amount of iron salt, a dinitrogen ligand, and a 1,2-dichloroalkane as oxidant to produce phenanthrene derivatives which are of interest in materials science. The authors pointed out that the reaction might proceed via two types of organoiron intermediates. Alkenyl iron species (A) generated through carbometalation of aryliron with alkynes to direct ortho-C–H bond ferration to generate seven-membered ring ferracycle intermediate is proposed. On the other hand, a five-membered ring ferracycle B formed by direct C–H activation to react with alkyne followed by reductive elimination upon interacting with dichloroalkane to deliver phenanthrene product is a preferred mechanism.

The above-discussed iron-catalyzed C–H bond functionalization using organometallic reagents all occurred on sp^2 hybridized carbon atoms. Nakamura and co-workers also reported iron-catalyzed arylation of $C(sp^3)$–H [18] of amines with an intramolecular trigger (Scheme 9.15). The authors reported that in the presence of an iron catalyst, 1-(2-iodobenzyl)pyrrolidine reacts with aryl Grignard reagents or aryl zinc reagents to generate C–H arylation at the C2-position on pyrrolidine with contemporary reduction of the C-I bond. Although the detailed mechanism of this reaction is still not clear, the reactivity and substrate scope clearly indicate this reaction proceeds through a distinct pathway compared with iron-catalyzed nitrogen directed $C(sp^2)$–H arylation discussed above. For elucidation of the reaction mechanism, the authors proposed that first a low-valent organoiron species interacts with aryl iodide to generate aryl radical (A) through single electron reduction of the aryl iodide. The aryl radical formed abstracts the C–H adjacent to nitrogen to generate a more stable α-aminoalkyl radical (B) via 1,5-hydrogen transfer. Finally, the alkyl radical rebinds to iron (C) and reductive elimination takes place to form carbon–carbon bond delivering the arylation product. The reaction utilized both the radical characteristic and the reactivity of

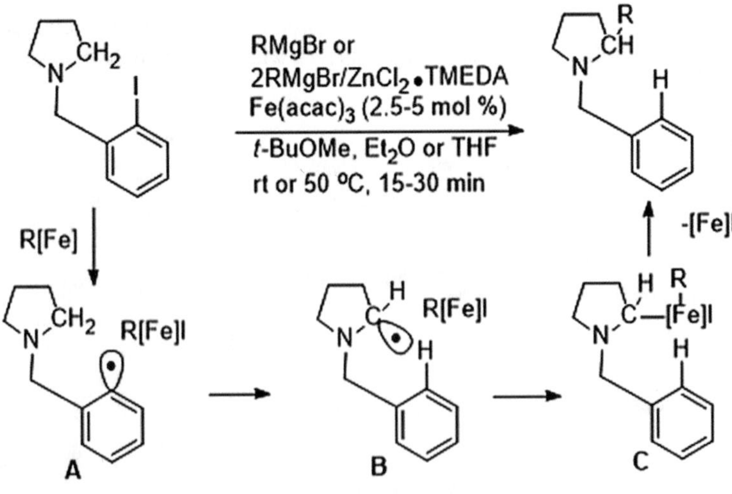

Scheme 9.15 Iron-catalyzed α-functionalization of aliphtic amine via $C(sp^3)$–H activation through 1,5-hydrogen transfer

organoiron intermediates as d-block organometallics. The development of this reaction provides some strategies for radical chemistry, such as radical translocation, which can be combined with organometallic reactivity of iron to develop new transformations.

The substrate scope of the iron-catalyzed 1,5-hydrogen transfer/C–H bond arylation reactions is very broad (Scheme 9.16). The reaction proceeded well not only for cyclic amine, but also for other tertiary acyclic amines. For example, when N-(2-iodobenzyl)-N-methylbutan-1-amine was used as substrate, C–H arylation took place on both the methyl and the α-C–H of the butyl chain. The arylation reaction on the α-C–H of the butyl chain (67%) is significantly faster than on methyl (8%), which is consistent with the stability of the generated alkyl radical intermediate. Besides aryl organometallic reagents, heteroaryl organometallic reagent, such as pyridin-3-ylmagnesium bromide, is also a suitable coupling partner. Alkenyl Grignard reagent and alkyl Grignard reagent can also be used to produce alkenylation and alkylation products.

The result of an isotope labeling experiments reported by Nakamura and co-workers confirms the intermolecular 1,5-hydrogen transfer mechanism (Scheme 9.17). When an 1:1 mixture of the piperidine substrates and the tetra-deuterated N,N-diethylamine was treated with 2.0 equivalents of 4-fluorophenyl magnesium bromides, two arylation products were generated quantitatively while the deuterium was only detected in the ortho-position of N,N-diethylamine products in 100% incorporation without any H-D scrambling.

Recently, Nakamura et al. reported an interesting example of iron-catalyzed arylation of allylic C–H bond with aryl Grignard reagents [19] (Scheme 9.18). In this reaction, the low-valent iron species generated by reduction with Grignard

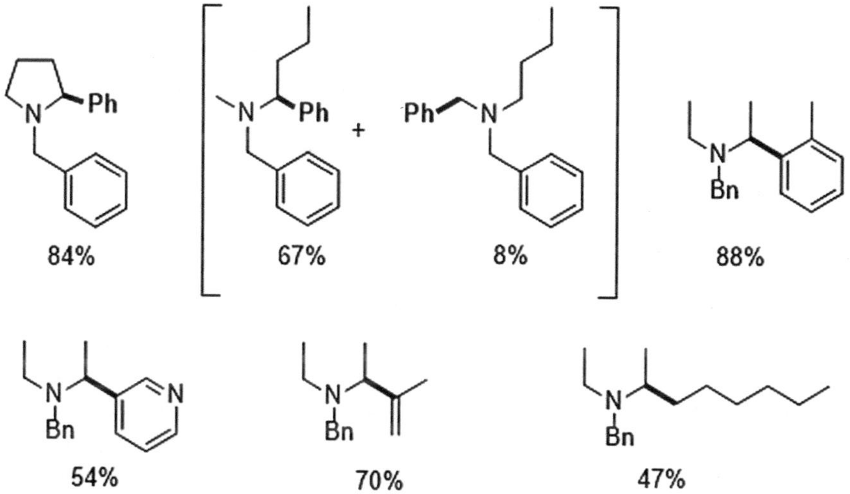

Scheme 9.16 Scope of iron-catalyzed α-functionalization of aliphtic amine via C(sp³)–H activation

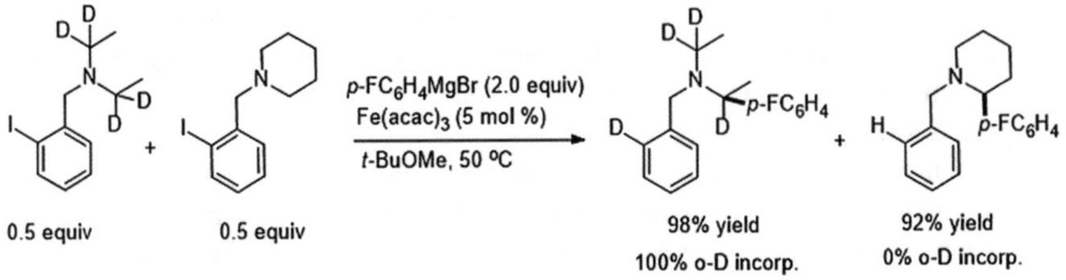

Scheme 9.17 Isotope labeling experiment for iron-catalyzed α-functionalization of aliphtic amine via C(sp^3)–H activation

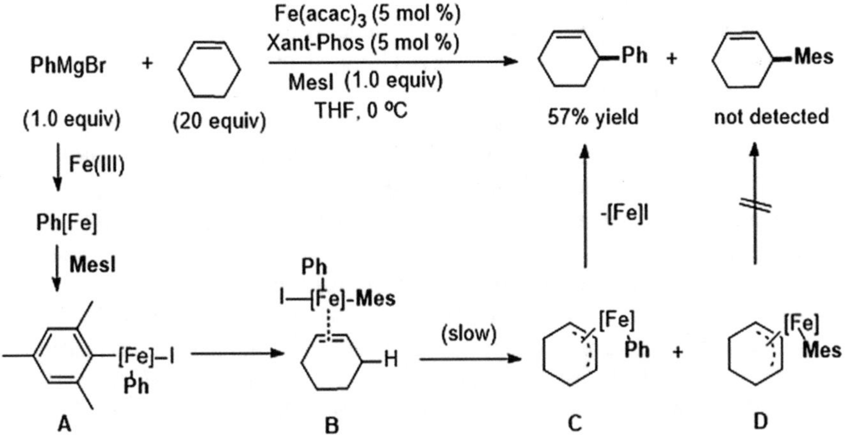

Scheme 9.18 Iron-catalyzed arylation of cyclohexene with PhMgBr in the presence of mesityl iodides

reagent activated C–H bond at the allylic position, and formed a new C-aryl bond through reductive elimination on iron. A catalyst composed of Fe(acac)$_3$ and 4,5-bis (diphenylphosphino)-9,9-dimethylxanthene was selected as the optimal catalyst. 2,4,6-Trimethyl-iodobenzene was found to be an efficient oxidant for this reaction. However, in this reaction, a large excess of allylic substrates has to be used and the efficiency of this reaction is low due to the undesired consumption of Grignard reagent by deportonation and homocouplings.

From the examples discussed above, it is clear that compared with ruthenium and palladium catalysis, the catalytic reactivity of iron is to a large extent still unknown, especially the catalytic reactivity of low-valent iron species has not been well exploited yet. The mechanisms of discovered reactions are mainly unclear. For iron-catalyzed C–H activation and C–C bond formation reactions, great opportunities are still in front of us. Considering the significance and impact of iron-catalysis, as well as the lack of precedent studies, there will be a great opportunity for synthetic chemists, especially the new generation to make new discoveries. From the examples in literatures, we may find most of the iron-catalyzed functionalizations of saturated C–H bonds proceed through iron participated radical pathways, and in low-valent iron-catalyzed C–H activation reactions, Grignard reagent and Girgnard-derived organozinc reagents have to be used as the coupling partners. With these problems in mind, we discovered a new facet of catalytic reactivity of organoiron species, that is, to catalyze a direct $C(sp^3)$–H functionalization through the formation of a $C(sp^3)$ ferracycle intermediate, which is a type of intermediate widely utilized in precious metal catalyzed C–H functionalizations. In Chap. 11 of this thesis, we disclosed a novel and practical iron-catalyzed C–H functionalization that uses commercially available organoboronate as coupling partner to create new C–C bond. The mechanism of this reaction has been studied and the key enabling factors for the successful utilization of a broad scope of boronate reagents were disclosed. The reaction scope is unprecedentedly broad, even rivaling the reported similar transformations catalyzed by precious metals. Thus, it reveals iron-catalyzed C–H activation process is not only an interesting scientific issue, but already showed high practicality. We also demonstrated in this work that iron-catalyzed C–H activation is not an alternative method for precious metal catalysis, but it can be used to solve the difficult problems in precious metal catalysis, such as to construct a conjugated diene and triene through C–H activation.

References

1. (a) Godula, K., & Sames, D. (2006). *Science, 312*, 67–72. (b) Lyons, T. W., & Sanford, M. S. (2010). *Chemical Reviews, 110*, 1147–1169. (c) Chem, X, Engle, K. M., Wang, D.-H., & Yu, J.-Q. (2009). *Angewandte Chemie International Edition, 48*, 5094–5115.
2. (a) Nakamura, E., & Yoshikai, N. (2010). The *Journal of Organic chemistry, 2010, 75*, 6061–6067. (b) Bolm, C. (2009). *Nauture Chemistry, 1*, 420. (c) Nakamura, E., & Sato, K. *Nature Materials, 10*, 158–161.
3. For a review on iron-catalyzed C–H activation, see: Sun, C.-L., Li, B.-J., & Shi, Z.-J. (2011). *Chemical Reviews, 111*, 1293–1314.
4. Vallee, F., Mousseau, J. J., & Charette, A. B. (2010). *Journal of the American Chemical Society, 132*, 1514–1516.
5. Liu, W., Cao, H., & Lei, A. (2010). *Angewandte Chemie International Edition, 49*, 2004–2008.
6. Sun, C.-L., Li H., Yu, D.-G., Yu, M., Zhou, X., Lu, X.-Y., Huang, K., Zheng, S.-F., Li, B.-J., & Shi, Z.-J. (2010). *Nauture Chemistry, 2*, 1044–1049.
7. Liu, W., Cao, H., Zhang, H., Zhang, H., Chung, K. H., He, C., Wang, H., Kwong, F. Y., & Lei, A. (2010). *Journal of the American Chemical Society, 132*,16737–16740.
8. Shirakawa, E., Itoh, K., Higashino, T., & Hayashi, T. (2010). *Journal of the American Chemical Society, 132*, 15537–15539.
9. Norinder, J., Matsumoto, A., Yoshikai, N., & Nakamura, E. (2008). *Journal of the American Chemical Society, 130*, 5858–5859.
10. Yoshikai, N., Matsumoto, A., Norinder, J., & Nakamura, E. (2009). *Angewandte Chemie International Edition, 48*, 2925–2928.

11. Tran, L. D., & Daugulis, O. (2010). *Organic Letters, 12*, 4277–4279.
12. (a) Ito S., Fujiwara Y., Nakamura E., & Nakamura M. (2009). *Organic Letters, 11*, 4306–4309. (b) Nakamura M., Matsuo, K. Ito S., & Nakamura E. (2004). *Journal of the American Chemical Society*, 126, 3686-3687.
13. Wen, J., Zhang, J., Chen, S.-Y., Li, J., & Yu, X.-Q. (2008). *Angewandte Chemie International Edition, 47*, 8897–8900.
14. Wen, J., Qin, S., Ma, L.-F., Dong, L., Zhang, J., Liu, S.-S., Duan, Y.-S., Chen, S.-Y., Hu, C.-W., & Yu, X.-Q. (2010). *Organic Letters, 12*, 2694–2697.
15. Ilies, L., Asako, S., & Nakamura, E. (2011). *Journal of the American Chemical Society, 133*, 7672–7675.
16. Ilies, L., Okabe, J., Yoshikai, N., & Nakamura, E. (2010). *Organic Letters, 12*, 2838–2840.
17. Matsumoto, A., Ilies, L., & Nakamura, E. (2011). *Journal of the American Chemical Society, 133*, 6557–6559.
18. Yoshikai, N., Mieczkowski, A., Matsumoto, A., Ilies, L., & Nakamura, E. (2010). *Journal of the American Chemical Society, 132*, 5568–5569.
19. Sekine, M., Ilies, L., & Nakamura, E. (2013). *Organic Letters, 15*, 714–717.

Chapter 10
β-Arylation of Carboxamides Via Iron-Catalyzed C(sp³)–H Bond Activation

Abstract Directed $C(sp^3)$–H bond functionalization has been studied mainly by using precious metal catalysts, such as Pd, Ru, Rh, and Ir under harsh conditions. Generally, these metal-catalyzed C–H functionalization reactions are based on the formation of a $C(sp^3)$-metallacycle. Iron-catalyzed $C(sp^3)$–H functionalization has been studied mainly using radical processes. Functionalization of an unactivated $C(sp^3)$–H bond via formation of a ferracycle intermediate is limited to stoichiometric reactions. We report here an iron/biphosphine-catalyzed directed arylation of a $C(sp^3)$–H bond in an aliphatic carboxamide with an organozinc reagent in high yield under mild oxidative conditions. The choice of the directing group and of the biphosphine ligand was crucial for the success of this reaction. This reaction shows selectivity for a primary C–H over a secondary one and is sensitive to steric factors on both the amide and the Grignard reagent. Various β-arylated aliphatic carboxamides can be readily prepared by using this method.

dppBz 85% (1.40 g)   **MeO-dppBz** 96%

10.1 Introduction

Because of the potential economic and environmental merits [1] compared with precious metals, iron catalysis [2] for $C(sp^2)$–H bond activation [3] to create a C–C bond has recently seen tremendous development [4], and iron catalysis for $C(sp^3)$–H activation [5, 6] is expanding as well [7, 8]. Except for a few examples, these

© Springer Nature Singapore Pte Ltd. 2017
R. Shang, *New Carbon–Carbon Coupling Reactions Based on Decarboxylation and Iron-Catalyzed C–H Activation*, Springer Theses,
DOI 10.1007/978-981-10-3193-9_10

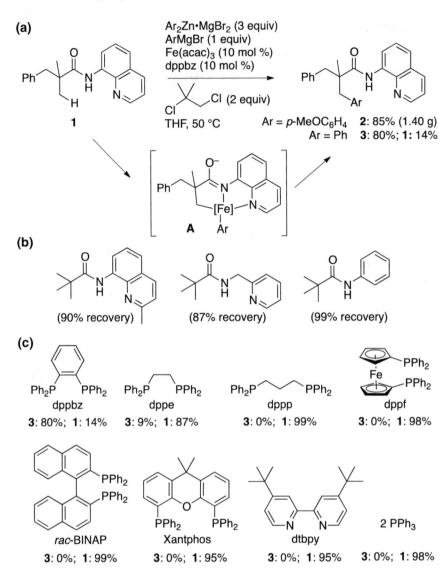

Fig. 10.1 Iron-catalyzed arylation of the β-methyl group of 2,2-disubstituted propionamide. **a** Representative example of conversion of **1** to **2** or **3** and a possible intermediate **A**. **b** Representative unreactive substrates under the conditions shown in (**a**). The recovery of the starting material is shown in *parentheses*. **c** Representative ligands examined as an illustration of the unique effectiveness of dppbz. The yield of phenylated product **3** and the recovery of **1** are shown. Reprinted with the permission from J. Am. Chem. Soc. **2013**, 135, 6030. Copyright 2011 American Chemical Society

reactions may be categorized as remote functionalization of the C–H bond, where an organometallic intermediate is stabilized by chelation to the nearby directing group (e.g., chelated metal homoenolate A in Fig. 10.1a). Having been interested in homoenolate chemistry for sometime [9], we conjectured that A in Fig. 10.1a may serve as a viable intermediate for the conversion of an aliphatic acid derivative, such as carboxamide 1 to a β-functionalized product (2 or 3). We report here an iron-catalyzed arylation of the β-methyl position of a 2,2-disubstituted propionamide bearing an 8-aminoquinolinyl group (NH-Q) [6a, 10], as the amide moiety in the presence of an organic oxidant [11] under mild thermal conditions. The reaction has less of the radical character previously observed in iron catalysis [7] and a more organometallic character [12], as the reaction is sensitive to the choice of the ligand and shows a complete preference for C–H bond activation on the methyl group over the benzyl group of 1(Fig. 10.1a).

10.2 Results and Discussion

10.2.1 Investigation of the Reaction Conditions

A typical procedure optimized after considerable experimentation is described first (Fig. 10.1a). The NH-Q amide 1(1.22 g, 4 mmol) was added to a tetrahydrofuran (THF) solution of freshly prepared ArMgBr (Ar = p-anisyl) (7 equiv) and $ZnBr_2 \cdot$ TMEDA (3 equiv). A solution of $Fe(acac)_3$ (10 mol%) and 1,2-bis (diphenylphosphino)benzene (dppbz) (10 mol%) in THF and 1,2-dichloroisobutane [13] (DCIB) (2 equiv) were added, and the mixture was heated at 50 °C for 36 h. Aqueous workup followed by column chromatography gave 1.40 g (85% yield) of arylated product 2 (Fig. 10.1a) together with the recovery of 1. Out of the 7 equiv of ArMgBr, 6 equiv forms 3 equiv of Ar_2Zn and 1 equiv deprotonates the amide proton. A small amount of the Ar group must have been consumed upon reaction with $Fe(acac)_3$. The use of a smaller amount of organometallic reagent resulted in a lower yield (45% with 2 equiv of organozinc reagent) and slower reaction. Omission of the zinc salt resulted in no formation of the desired product. For reasons yet to be probed, increasing the amounts of the zinc reagent and the catalyst and using a longer reaction time did not result in higher conversion. Addition of a catalytic amount of water or the use of old Grignard reagent significantly lowered the reaction yield, suggesting that the presence of alkoxide impedes the reaction.

Under similar conditions, the reaction of phenylmagnesium bromide gave 3 in 80% yield together with recovered 1 (14%) and biaryl (15% based on PhMgBr) resulting from iron-mediated homocoupling [14]. Interestingly, most of the biaryl was formed after the product formation stopped. We found no products from either arylation at the benzylic position 7a of 1, further reaction of the product 2 or 3, or arylation of the carboxamide nitrogen [15]. This demonstrates the highly regional and chemical selectivity of the reaction.

The NH-Q directing group and the dppbz ligand were found to be uniquely effective for the reaction (Fig. 10.1). For instance, a 2-methyl group on the quinoline entirely stopped the reaction, and a 2-picoline analog and a simple N-phenylcarboxamide did not take part in the reaction at all (Fig. 10.1b). The

N-methylated derivative of 4 did not react at all (Fig. 10.1b). The importance of the ligand is illustrated in Fig. 10.1c. The reaction did not proceed at all in the absence of a ligand. The bidentate ligand 1,2-bis(diphenylphosphino)-ethane (dppe), which is similar to dppbz except for its slightly larger bite angle and a more flexible backbone, gave the product in 9% yield. Other bidentate phosphine ligands with larger bite angles and various degrees of flexibility were entirely inefficient.

Bipyridine-type ligands that are the ligand of choice for iron-catalyzed C(sp²)–H bond activation were ineffective, and monophosphine ligands such as PPh₃ were also ineffective. Such high sensitivity to the ligand structure is less consistent with either a pure radical mechanism or sole inclusion of organozinc species than with a chelated iron intermediate [16, 17] such as A.

10.2.2   Exploration of the Substrate Scope

We found that except for the structures of the directing group and the ligand, the reaction was also sensitive to the structure of the substrate, as summarized in Table 10.1, at the bottom of which unreactive substrates are listed. Carboxamides possessing 3-aryl- and 3-naphthyl-2,2-dimethylpropionamide reacted exclusively on one of the two methyl groups (entries 1–6) with retention of fluorine, chlorine, and bromine groups. Pivalamide 4 (entry 7) gave a mixture of monoarylated and diarylated products (1 and 3), and no further arylation of 3 occurred. Replacement of one methyl group in the pivalamide with an ethyl group (entry 8) resulted in selective monoarylation, but replacement with a phenyl group shut off the reaction (bottom of Table 10.1). Cyclohexanecarboxamide 5 (entry 9) and cyclopentanecarboxamide 6 (entry 10) reacted well, whereas the corresponding cyclobutane- and cyclopropanecarboxamides 7 and 8 did not give the desired product at all. One key feature that might be considered crucial for efficient C–H activation may be the CH₃–C–C(O) bond angle (θ in Fig. 10.2): this angle is much wider for the unreactive substrates 7 and 8 than for 5 and 6, making the distance between the β-H and amide nitrogen (l) longer and thus the formation of a chelate intermediate A less feasible. However, a smaller θ may not be sufficient, as most of the reactive substrates, such as 1 and 4 as well as unreactive substrates including 2, 3, and propionamide (Table 10.1 bottom) have $\theta \approx 107$–$109°$ (data not shown). We note that cyclopropanecarboxamide 8 was completely recovered, and a ring-opened product was not produced at all. Cyclohexanecarboxamides not possessing the α-methyl group did not give the desired product, as shown at the bottom of Table 10.1.

Para-substituted arylzinc reagents (Table 10.1, entries 12, 13, 16, and 18) reacted well, and meta-substitution (entries 14 and 17) resulted in satisfactory yields, while ortho substitution totally shut off the reaction (entry 15). Electron-deficient organometallic reagents (entries 16 and 17) tended to give lower yields than electron-rich reagents (entries 12, 13, and 18). A 2-naphthylzinc reagent also gave a satisfactory yield (entry 19). Alkyl- and alkenylzinc reagents did not react under these reaction conditions.

Table 10.1 Iron-catalyzed arylation of 2,2-disubstituted propionamides with organozinc reagent[a]

entry	substrate[b]	product[b]		yield (%)[c]
1		**2** X = H		79
2		X = F		74
3		X = Cl		71
4		p-MeOC$_6$H$_4$ X = Br		73
5				71
6				83
7[d]	**4**	**1** Ph **3**		33+31
8				53
9	**5**			75
10	**6**			69
11		**2** Ar = Ph		80
12		Ar = p-Tol		78
13		Ar = p-t-BuC$_6$H$_4$		85
14		Ar = m-Tol		78
15		Ar = o-Tol		0
16		Ar = p-FC$_6$H$_4$		56
17		Ar = m-FC$_6$H$_4$		51
18		Ar = p-Me$_2$NC$_6$H$_4$		80
19		Ar = 2-Naphthyl		69

(substrates at bottom: NH-Q; NH-Q; Ph—NH-Q; NH-Q **7**; NH-Q **8**)

[a]The reaction was performed under the conditions in Fig. 10.1a using 0.5 mmol of substrate. Unreactive substrates (<5% yield) are shown at the bottom. [b]Q = 8-quinolinyl. [c]Determined by isolation. [d]Determined by GC in the presence of tridecane as an internal standard. [e]20 mol% of catalyst was used

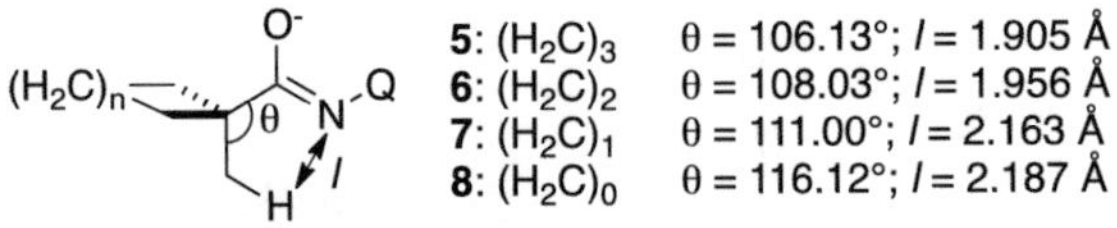

5: (H$_2$C)$_3$ θ = 106.13°; l = 1.905 Å
6: (H$_2$C)$_2$ θ = 108.03°; l = 1.956 Å
7: (H$_2$C)$_1$ θ = 111.00°; l = 2.163 Å
8: (H$_2$C)$_0$ θ = 116.12°; l = 2.187 Å

Fig. 10.2 Bond angle and atomic distance for cycloalkylcarboxamides 5–8. MMFF-optimized with H–C–C–C = N fixed in the plane

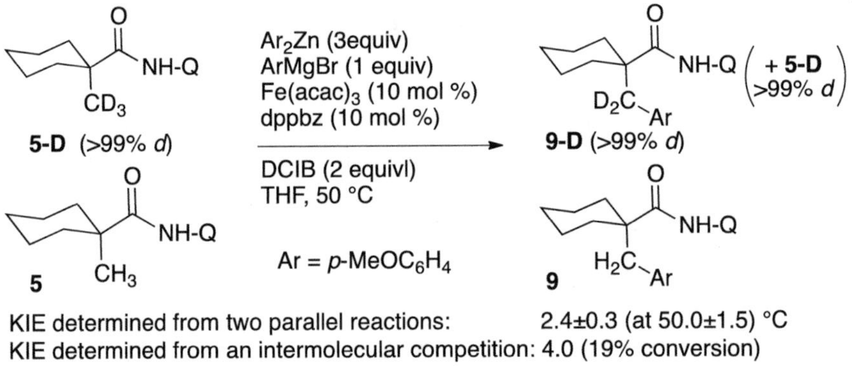

KIE determined from two parallel reactions: 2.4±0.3 (at 50.0±1.5) °C
KIE determined from an intermolecular competition: 4.0 (19% conversion)

Scheme 10.1 KIE experiments. Reprinted with the permission from J. Am. Chem. Soc. **2013**, 135, 6030. Copyright 2011 American Chemical Society

10.2.3 Deuterated Experiments

Kinetic isotope effect (KIE) experiments indicated that the cleavage of the C–H bond is the rate-determining step of the reaction. As depicted in Scheme 10.1, competition experiments between 5 and the deuterated substrate 5-D showed a primary KIE of 2.4 when the reactions were performed in parallel and an inter-molecular KIE of 4.0 (at 19% conversion). The organometallic reagent takes up the β-hydrogen, as demonstrated by partial deuterium incorporation into the recovered organometallic reagent for the reaction of 5-D. Deuterium scrambling in the product 9-D or the recovered 5-D was not observed.

10.3 Conclusion

In summary, we report the Fe-catalyzed arylation reaction of replacing a $C(sp^3)$–H bond with a new C-aryl bond at the β-position of a 2,2-disubstituted carboxamide, where the quinolinamide group acts as a uniquely effective directing group. The overwhelmingly higher reactivity of a methyl group over a benzylic group excludes a radical mechanism, and the high sensitivity of the yield to the structure of the substrate and the ligand suggests involvement of organoiron intermediates in some crucial steps. Further understanding of the reaction parameters in the present reaction will uncover guidelines for designing efficient iron catalysts.

10.4 Experimental Section and Compound Data

10.4.1 General Information

All the reactions dealing with air- or moisture-sensitive compounds were carried out in a dry reaction vessel under a positive pressure of argon. Air- and moisture-sensitive liquids and solutions were transferred via syringe or Teflon cannula. Analytical thin-layer chromatography was performed using glass plates pre-coated with 0.25-mm 230–400 mesh silica gel impregnated with a fluorescent indicator (254 nm). Thin-layer chromatography plates were visualized by exposure to ultraviolet light (UV). Organic solutions were concentrated by rotary evaporation at ca. 15 Torr (evacuated with a diaphragm pump). Flash column chromatography was performed as described by Still et al. (J. Org. Chem. **1978**, 43, 2923–2924), employing Kanto Silica gel 60 (spherical, neutral, 140–325 mesh). Gas-liquid chromatographic (GLC) analysis was performed on a Shimadzu 14A or 14B machine equipped with glass capillary column HR-1 (0.25-mm i.d. × 25 m). Gel permeation column chromatography was performed on a Japan Analytical Industry LC-908 (eluent: toluene) with JAIGEL 1H and 2H polystyrene columns.

Proton nuclear magnetic resonance (1H NMR) and carbon nuclear magnetic resonance (13C NMR) spectra were recorded with JEOL ECA-500 NMR spectrometers. Chemical data for protons are reported in parts per million (ppm, δ scale) downfield from tetramethylsilane and are referenced internally to tetramethylsilane. Carbon nuclear magnetic resonance spectra (^{13}C NMR) were recorded at 125 MHz. Chemical data for carbons are reported in parts per million (ppm, δ scale) and are referenced to the carbon resonance of the solvent ($CDCl_3$: δ = 77.0). The data is presented as follows: chemical shift, multiplicity (s = singlet, d = doublet, t = triplet, q = quartet, m = multiplet and/or multiplet resonances, br = broad), coupling constant in Hertz (Hz), and integration. Mass spectra (GS MS) were taken at SHIMADZU Parvum 2 gas chromatograph mass spectrometer.

Commercial reagents were purchased from Tokyo Kasei Co., Aldrich Inc., and other commercial suppliers and were used after appropriate purification. Anhydrous ethereal solvents (stabilizer-free) were purchased from WAKO Pure Chemical and purified by a solvent purification system (GlassContour) equipped with columns of activated alumina and supported copper catalyst (Q-5) prior to use (Organometallics **1996**, 15, 1518–1520). The content of water in the solvents or the substrates were confirmed to be less than 30 ppm by Karl Fischer moisture titrator. The Grignard reagents were prepared from the corresponding bromide and magnesium turnings in anhydrous tetrahydrofuran (THF), and were titrated prior to use. Fe(acac)3 (99.9+ %) was purchased from Aldrich Inc. and used as received. Iron salts from other sources are mentioned in the text where appropriate. 1,2-bis(Diphenylphosphino) benzene (dppbz) was purchased from Aldrich Inc. and used as received. 1,2-Dichloroisobutane (DCIB) was purchased from Tokyo Kasei Co. and used as received. $ZnBr_2 \cdot$ TMEDA was prepared according to the literature procedure (*Chem. Lett.* **1977**, 679–682).

The geometries of cycloalkanecarboxamides **5–8** were optimized using the Spartan '06 package (Wavefunction, Inc. Irvine, CA) using the Merck Molecular Force Field (MMFF) (Thomas A. Halgren, J. Comp. Chem.; **1996**; 490–519), where the reactive H, the beta-C, the alpha-C and the C(O)N of the directing group were fixed in the same plane.

Caution We observed that contamination of the reaction mixture by atmospheric moisture significantly reduces the yield, and Grignard reagents stored for a long time gave lower yields than freshly prepared ones. We sometimes observed complete suppression of the reaction when using an old Grignard reagent or incompletely anhydrous conditions, and we ascribe this to the poisoning of the iron catalyst by hydroxide species. For best results, all the experiments described herein must be performed under rigorous anhydrous conditions with good quality (preferably freshly prepared) organometallic reagents.

10.4.2 Investigation of the Key Reaction Parameters

General Procedure for the Optimization Study

Table 10.2 shows a typical procedure for the optimization study: A solution of PhMgBr in THF (1.02 mol/L, 0.7 mmol) was slowly added to a mixture of *N*-(quinolin-8-yl) aliphatic amide (0.1 mmol) and $ZnBr_2 \cdot$ TMEDA (0.3 mmol) using a gastight syringe under argon. The resulted mixture was stirred at r.t. for 30 min, then a solution of Fe(acac)₃ (0.01 mmol) and 1,2-bis(diphenylphosphino)benzene (dppbz, 0.01 mmol) in 1.0 mL anhydrous THF was added. The resulted black mixture was stirred at r.t. for 10 min, and then 1,2-dichloroisobutane (0.2 mmol) was added. The mixture was stirred at 50 °C for 36 h. After cooling to r.t., tridecane (0.1 mmol) was added as an internal standard. Aqueous Rochelle's salt was then added, and the organic layer was extracted with diethyl ether and passed over a Florisil column. The mixture was analyzed by GC using tridecane as an internal standard.

Table 10.2 Optimization of the reaction conditions

modification from the optimal condition	yield of 3[a] (%)	recovery of 1[a] (%)
1. Optimal Condition	84	19 (15% Ph-Ph)[b]
2. ZnCl₂ instead of ZnBr₂•TMEDA	32	66
3. ZnCl₂•TMEDA instead of ZnBr₂•TMEDA	75	23
4. without Zinc Source	0	28
5. ZnCl₂•TMEDA and PhMgCl instead	0	99
6. PhMgCl instead	2	94
7. dppe instead of dppbz	9	87
8. dppp instead of dppbz	0	99
9. dppf instead of dppbz	0	98
10. rac-BINAP instead of dppbz	0	99
11. Xant-Phos instead of dppbz	0	95
12. Fe(acac)₂ instead of Fe(acac)₃	55	43
13. FeF₂ instead of Fe(acac)₃	trace	97
14. no Fe	0	99
15. 5.0 eq of PhMgBr and 2.0 eq of ZnBr₂•TMEDA	45	52
16. added 1μL water	34	65

[a]Based on **1**, estimated by GC using tridecane as an internal standard. [b]Based on PhMgBr

A solution of PhMgBr in THF (1.02 mol/L, 1.1 mmol) was slowly added using a gastight syringe to a mixture of *N*-(quinolin-8-yl)pivalamide (0.1 mmol) and ZnBr₂ · TMEDA (0.5 mmol) under argon (Scheme 10.2). The resulted mixture was stirred at r.t. for 30 min, then a solution of Fe(acac)₃ (0.1 mmol) and dppbz (0.1 mmol) in 2.0 mL anhydrous THF was added. The resulted black mixture was stirred at r.t. for 10 min, and then 1,2-dichloroisobutane (0.4 or 0 mmol) was added. The mixture was stirred at 50 °C for 24 h. After cooling to r.t., tridecane

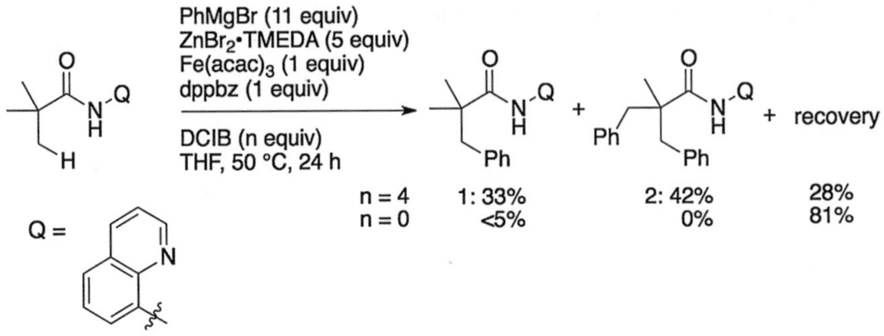

Scheme 10.2 Reaction with 1.0 equiv of iron and ligand

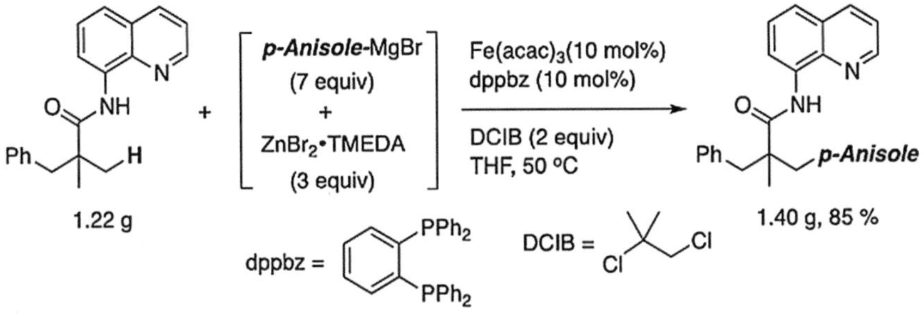

Scheme 10.3 Directed C–H arylation of 2,2-dimethyl-3-phenyl-N-(quinolin-8-yl)propanamide

(0.1 mmol) was added as an internal standard. Aqueous Rochelle's salt was then added, and the organic layer was extracted with diethylether and passed over a Florisil column. The mixture was analyzed by GC using tridecane as an internal standard.

Iron-Catalyzed Arylation of Aliphatic Carboxamide

Scheme 10.3 shows the directed C–H arylation of 2,2-dimethyl-3-phenyl-N-(quinolin-8-yl)propanamide on gram scale.

A freshly prepared solution of (4-methoxyphenyl)magnesium bromide in THF (0.5 mol/L, 28.0 mmol, 56 mL) was slowly added using a gastight syringe to a mixture of 2,2-dimethyl-3-phenyl-N-(quinolin-8-yl)propanamide (**1**, 4.0 mmol, 1.22 g) and ZnBr$_2$ · TMEDA (12.0 mmol, 4.08 g) under argon. The resulted mixture was stirred at r.t. for 30 min, then a solution of Fe(acac)$_3$ (0.4 mmol, 140 mg) and dppbz (0.4 mmol, 184 mg) in anhydrous THF (5.0 mL) was added. The resulted black mixture was stirred at r.t. for 10 min, then 1,2-dichloroisobutane (8 mmol, 920 μL) was added via syringe. The mixture was stirred at 50 °C for 36 h. After cooling to r.t., aqueous Rochelle's salt was added (20 mL), and the organic layer was extracted with diethyl ether (30 mL × 3). The combined organic layers were washed with brine, passed through a pad of Florisil, dried over MgSO$_4$, and concentrated under reduced pressure. The crude mixture was purified by column chromatography on silica gel (hexane:ethyl acetate = 10:1) to afford 2-benzyl-3-(4-methoxyphenyl)-2-methyl-N-(quinolin-8-yl)propanamide as a pale yellow oil (1.40 g, 85% yield).

General Procedure for Table 10.1

A freshly prepared solution of an arylmagnesium bromide in THF (7.0 equiv, 3.5 mmol, titrated before use) was slowly added using a gastight syringe to a mixture of the corresponding carboxamide (1.0 equiv, 0.50 mmol) and ZnBr$_2$ · TMEDA (3.0 equiv, 1.5 mmol, 510 mg) under argon. The resulted mixture was stirred at r.t. for 30 min, then a solution of Fe(acac)$_3$ (10 mol%, 0.05 mmol, 17 mg) and dppbz (10 mol%, 0.05 mmol, 23 mg) in anhydrous THF (2.0 mL) was added. The resulted black mixture was stirred at r.t. for 10 min. Then 1,2-dichloroisobutane (1.00 mmol, 115 uL) was added via syringe and the mixture was stirred at 50 °C for 36 h. After cooling to r.t., aqueous Rochelle's salt was added (5 mL), and the organic layer was extracted with diethyl ether (15 mL × 3). The combined organic layers were washed with brine, passed through a pad of Florisil, dried over MgSO$_4$, and concentrated under reduced pressure. The crude mixture was purified by column chromatography on silica gel (hexane:ethyl acetate = 20:1–10:1) to afford the corresponding arylated product.

10.4.3 Characterization of Compounds

2-Benzyl-3-(4-Methoxyphenyl)-2-Methyl-N-(Quinolin-8-yl)Propanamide

^{1}H NMR (500 MHz, CDCl3) δ 9.95 (s, 1H), 8.92 (d, J = 4.6 Hz, 1H), 8.66–8.54 (m, 1H), 8.05 (dd, J = 8.1, 2.4 Hz, 1H), 7.56 (t, J = 8.0 Hz, 1H), 7.46 (d, J = 8.2 Hz, 1H), 7.32 (dd, J = 8.2, 4.1 Hz, 1H), 7.25 (d, J = 7.1 Hz, 2H), 7.22–7.08 (m, 5H), 6.72 (d, J = 8.1 Hz, 2H), 3.66 (s, 3H), 3.49 (dd, J = 28.1, 13.2 Hz, 2H), 2.78 (dd, J = 19.0, 13.5 Hz, 2H), 1.35 (s, 3H).

^{13}C NMR (126 MHz, CDCl3) δ 174.55, 158.07, 147.93, 138.54, 137.63, 135.92, 134.14, 131.18, 130.26, 129.52, 127.90, 127.66, 127.20, 126.27, 121.29, 116.12, 113.31, 54.92, 49.96, 46.40, 45.74, 19.54 (two aromatic carbon signals are overlapping).

GC MS (EI) *m/z* (relative intensity): 410 (M+, 2), 319 (61), 175 (100), 171 (40), 144 (30), 121 (91), 91 (48), 77 (11).

2-Benzyl-3-(4-Methoxyphenyl)-2-Methyl-*N*-(Quinolin-8-yl)Propanamide obtained in 79% yield as a white viscous liquid.

¹H NMR (500 MHz, CDCl3) δ 9.95 (s, 1H), 8.92 (d, J = 4.6 Hz, 1H), 8.66–8.54 (m, 1H), 8.05 (dd, J = 8.1, 2.4 Hz, 1H), 7.56 (t, J = 8.0 Hz, 1H), 7.46 (d, J = 8.2 Hz, 1H), 7.32 (dd, J = 8.2, 4.1 Hz, 1H), 7.25 (d, J = 7.1 Hz, 2H), 7.22–7.08 (m, 5H), 6.72 (d, J = 8.1 Hz, 2H), 3.66 (s, 3H), 3.49 (dd, J = 28.1, 13.2 Hz, 2H), 2.78 (dd, J = 19.0, 13.5 Hz, 2H), 1.35 (s, 3H).

¹³C NMR (126 MHz, CDCl3) δ 174.55, 158.07, 147.93, 138.54, 137.63, 135.92, 134.14, 131.18, 130.26, 129.52, 127.90, 127.66, 127.20, 126.27, 121.29, 116.12, 113.31, 54.92, 49.96, 46.40, 45.74, 19.54 (two aromatic carbon signals are overlapping).

GC MS (EI) *m/z* (relative intensity): 410 (M+, 2), 319 (61), 175 (100), 171 (40), 144 (30), 121 (91), 91 (48), 77 (11).

2-(4-Fluorobenzyl)-3-(4-Methoxyphenyl)-2-Methyl-*N*-(Quinolin-8-yl) Propanamide obtained in 74% yield as a white viscous liquid.

¹H NMR (500 MHz, CDCl3) δ 9.91 (s, 1H), 8.87 (dd, J = 7.6, 1.3 Hz, 1H), 8.61 (dd, J = 4.2, 1.7 Hz, 1H), 8.08 (dd, J = 8.3, 1.6 Hz, 1H), 7.55 (t, J = 8.0 Hz, 1H), 7.47 (dd, J = 8.3, 1.2 Hz, 1H), 7.34 (dd, J = 8.2, 4.2 Hz, 1H), 7.16 (dd, J = 8.5, 5.5 Hz, 2H), 7.13 (d, J = 8.5 Hz, 2H), 6.84 (t, J = 8.7 Hz, 2H), 6.71 (d, J = 8.5 Hz, 2H), 3.66 (s, 3H), 3.44 (dd, J = 23.2, 13.4 Hz, 2H), 2.72 (dd, J = 13.7, 13.7 Hz, 2H), 1.31 (s, 3H).

¹³C NMR (126 MHz, CDCl3) δ 174.39, 161.57 (d, J_F = 244.3 Hz), 158.14, 148.03, 138.57, 136.02, 134.04, 133.34 (d, J_F = 3.3 Hz), 131.61 (d, J_F = 7.8 Hz),

131.21, 129.40, 127.73, 127.22, 121.46, 121.39, 116.19, 114.71 (d, JF = 21.0 Hz), 113.37, 54.97, 50.00, 45.79, 45.53, 19.50.

GC MS (EI) m/z (relative intensity): 428 (M+, 2), 319 (56), 175 (100), 144 (29), 121 (83), 109 (38).

2-(4-Chlorobenzyl)-3-(4-Methoxyphenyl)-2-Methyl-*N*-(Quinolin-8-yl) Propanamide obtained in 71% yield as a white viscous liquid.

^{1}H NMR (500 MHz, CDCl3) δ 9.89 (s, 1H), 8.85 (dd, J = 7.6, 1.2 Hz, 1H), 8.62 (dd, J = 4.2, 1.6 Hz, 1H), 8.08 (dd, J = 8.2, 1.5 Hz, 1H), 7.55 (t, J = 7.9 Hz, 1H), 7.48 (dd, J = 8.3, 1.2 Hz, 1H), 7.36 (dd, J = 8.2, 4.2 Hz, 1H), 7.15–7.09 (m, 6H), 6.70 (d, J = 8.5 Hz, 2H), 3.66 (s, 3H), 3.43 (dd, J = 29.6, 13.4 Hz, 2H), 2.71 (dd, J = 24.0, 13.0 Hz, 2H), 1.30 (s, 3H).

^{13}C NMR (126 MHz, CDCl3) δ 174.26, 158.16, 148.06, 138.57, 136.20, 136.03, 134.03, 132.16, 131.55, 131.22, 129.31, 128.07, 127.74, 127.22, 121.49, 121.42, 116.21, 113.38, 55.00, 49.95, 45.85, 45.61, 19.52.

GC MS (EI) m/z (relative intensity): 444 (M+, 1), 319 (56), 175 (100), 144 (127), 121 (75), 91 (8).

2-(4-Bromobenzyl)-3-(4-Methoxyphenyl)-2-Methyl-*N*-(Quinolin-8-yl) Propanamide obtained in 73% yield as a pale yellow viscous liquid.

^{1}H NMR (500 MHz, CDCl3) δ 9.90 (s, 1H), 8.87 (dd, J = 8.9, 7.9 Hz, 1H), 8.61 (dd, J = 4.2, 1.6 Hz, 1H), 8.06 (dd, J = 8.2, 1.5 Hz, 1H), 7.55 (t, J = 8.0 Hz, 1H), 7.47 (d, J = 8.2 Hz, 1H), 7.34 (dd, J = 8.2, 4.2 Hz, 1H), 7.26 (d, J = 8.4 Hz, 2H), 7.13 (d, J = 8.6 Hz, 2H), 7.07 (d, J = 8.3 Hz, 2H), 6.71 (d, J = 8.7 Hz, 2H), 3.65 (s, 3H), 3.43 (dd, J = 22.7, 13.3 Hz, 2H), 2.71 (dd, J = 33.0, 13.3 Hz, 2H), 1.31 (s, 3H).

^{13}C NMR (126 MHz, CDCl3) δ 174.19, 158.13, 148.03, 138.51, 136.68, 135.97, 133.97, 131.92, 131.18, 130.98, 129.24, 127.69, 127.16, 121.47, 121.38, 120.31, 116.17, 113.36, 54.95, 49.86, 45.82, 45.62, 19.49.

GC MS (EI) m/z (relative intensity): 489 (M+, 1), 488 (M+, 1), 320 (14), 319 (61), 175 (100), 144 (29), 121 (70), 191 (10).

2-(3-Methoxybenzyl)-3-(4-Methoxyphenyl)-2-Methyl-*N*-(Quinolin-8-yl) Propanamide obtained in 71% yield as a pale white viscous liquid.

^{1}H NMR (500 MHz, CDCl3) δ 9.93 (s, 1H), 8.87 (dd, J = 7.6, 1.1 Hz, 1H), 8.62 (dd, J = 4.2, 1.6 Hz, 1H), 8.09 (dd, J = 8.3, 1.6 Hz, 1H), 7.54 (t, J = 8.0 Hz, 1H), 7.47 (dd, J = 8.3, 1.2 Hz, 1H), 7.36 (dd, J = 8.2, 4.2 Hz, 1H), 7.12 (d, J = 8.6 Hz, 2H), 7.07 (t, J = 7.9 Hz, 1H), 6.79 (d, J = 7.3 Hz, 1H), 6.75 (s, 1H), 6.70 (d, J = 8.5 Hz, 2H), 6.64 (dd, J = 8.2, 2.5 Hz, 1H), 3.67 (s, 3H), 3.55 (s, 3H), 3.44 (dd, J = 21.2, 13.3 Hz, 2H), 2.73 (dd, J = 13.3, 6.6 Hz, 2H), 1.31 (s, 3H).

^{13}C NMR (126 MHz, CDCl3) δ 174.75, 159.17, 158.11, 148.04, 139.22, 138.63, 136.03, 134.17, 131.25, 129.58, 128.83, 127.74, 127.25, 122.76, 121.42, 121.40, 116.23, 115.26, 113.37, 112.33, 55.02, 54.85, 50.03, 46.58, 45.82, 19.60.

GC MS (EI) m/z (relative intensity): 440 (M+, 2), 319 (58), 175 (100), 144 (28), 121 (98), 91 (19), 77 (12).

2-Benzyl-2-Methyl-3-Phenyl-*N*-(Quinolin-8-yl)Propanamide

^{1}H NMR (500 MHz, CDCl3) δ 9.93 (s, 1H), 8.90 (dd, J = 7.7, 1.2 Hz, 1H), 8.61 (dd, J = 4.2, 1.6 Hz, 1H), 8.08 (dd, J = 8.2, 1.5 Hz, 1H), 7.56 (t, J = 8.0 Hz, 1H), 7.48 (dd, J = 8.3, 1.2 Hz, 1H), 7.34 (dd, J = 8.2, 4.2 Hz, 1H), 7.22 (d, J = 8.0 Hz, 4H), 7.17 (dd, J = 8.1, 6.8 Hz, 4H), 7.12 (ddd, J = 8.3, 2.5, 1.2 Hz, 2H), 3.51 (d, J = 13.2 Hz, 2H), 2.80 (d, J = 13.2 Hz, 2H), 1.33 (s, 3H).

^{13}C NMR (126 MHz, CDCl3) δ 174.52, 148.00, 138.63, 137.59, 136.01, 134.17, 130.32, 127.96, 127.73, 127.29, 126.34, 121.36, 116.21, 49.91, 46.56, 19.59(two aromatic carbon signals are overlapping).

GC MS (EI) *m/z* (relative intensity): 380 (M+, 4), 289 (87), 171 (100), 145 (73), 117 (27), 91 (94).

1-(4-Methoxybenzyl)-*N*-(Quinolin-8-yl)Cyclohexanecarboxamide the crude mixture was purified by column chromatography on silica gel (*n*-hexane: EtOAc = 15:1) to afford the titled compound in 75% yield as a white liquid.

^{1}H NMR (500 MHz, CDCl3) δ 10.02 (s, 1H), 8.82 (dd, J = 7.6, 1.3 Hz, 1H), 8.68 (dd, J = 4.2, 1.6 Hz, 1H), 8.10 (dd, J = 8.3, 1.6 Hz, 1H), 7.53 (t, J = 7.9 Hz, 1H), 7.46 (dd, J = 8.3, 1.2 Hz, 1H), 7.38 (dd, J = 8.2, 4.2 Hz, 1H), 7.01 (d, J = 9.0 Hz, 2H), 6.62 (d, J = 8.5 Hz, 2H), 3.60 (s, 3H), 2.91 (s, 2H), 2.29–2.20 (m, 2H), 1.80–1.66 (m, 2H), 1.64–1.45 (m, 5H), 1.38–1.24 (m, 1H).

^{13}C NMR (126 MHz, CDCl3) δ 174.61, 158.04, 147.98, 138.63, 136.00, 134.41, 130.92, 129.04, 127.74, 127.30, 121.31, 121.04, 116.12, 113.15, 54.91, 49.55, 34.07, 25.86, 23.06 (two aliphatic carbon signals are overlapping).

GC MS (EI) *m/z* (relative intensity): 374 (M+, 14), 206 (17), 171 (59), 144 (83), 121 (100), 109 (10), 77 (10).

1-(4-Methoxybenzyl)-*N*-(Quinolin-8-yl)Cyclopentanecarboxamide obtained in 69% yield as a white oil.

^{1}H NMR (500 MHz, CDCl3) δ 9.98 (s, 1H), 8.78 (dd, J = 7.6, 1.3 Hz, 1H), 8.71 (dd, J = 4.2, 1.6 Hz, 1H), 8.12 (dd, J = 8.2, 1.6 Hz, 1H), 7.53 (t, J = 7.9 Hz, 1H), 7.47 (dd, J = 8.2, 1.3 Hz, 1H), 7.40 (dd, J = 8.2, 4.2 Hz, 1H), 7.07 (d, J = 8.6 Hz, 2H), 6.65 (d, J = 8.6 Hz, 2H), 3.64 (s, 3H), 3.05 (s, 2H), 2.29–2.18 (m, 2H), 1.89–1.75 (m, 6H).

^{13}C NMR (126 MHz, CDCl3) δ 175.63, 158.09, 148.04, 138.62, 136.10, 134.54, 130.61, 130.38, 127.82, 127.37, 121.40, 121.08, 116.09, 113.42, 57.73, 55.00, 43.80, 35.23, 23.94.

GC MS (EI) *m/z* (relative intensity): 360 (M+, 14), 239 (18), 188 (23), 171 (51), 144 (100), 121 (95), 95 (11), 77 (12).

2-Benzyl-2-Methyl-3-Phenyl-*N*-(Quinolin-8-yl)Propanamide the crude mixture was purified by gel permeation column chromatography (toluene) to afford the title compound in 80% yield as a pale white viscous liquid.

^{1}H NMR (500 MHz, CDCl3) δ 9.93 (s, 1H), 8.90 (dd, J = 7.7, 1.2 Hz, 1H), 8.61 (dd, J = 4.2, 1.6 Hz, 1H), 8.08 (dd, J = 8.2, 1.5 Hz, 1H), 7.56 (t, J = 8.0 Hz, 1H), 7.48 (dd, J = 8.3, 1.2 Hz, 1H), 7.34 (dd, J = 8.2, 4.2 Hz, 1H), 7.22 (d, J = 8.0 Hz, 4H), 7.17 (dd, J = 8.1, 6.8 Hz, 4H), 7.12 (ddd, J = 8.3, 2.5, 1.2 Hz, 2H), 3.51 (d, J = 13.2 Hz, 2H), 2.80 (d, J = 13.2 Hz, 2H), 1.33 (s, 3H).

^{13}C NMR (126 MHz, CDCl3) δ 174.52, 148.00, 138.63, 137.59, 136.01, 134.17, 130.32, 127.96, 127.73, 127.29, 126.34, 121.36, 116.21, 49.91, 46.56, 19.59.

GC MS (EI) *m/z* (relative intensity): 380 (M+, 4), 289 (87), 171 (100), 145 (73), 117 (27), 91 (94).

2-Benzyl-3-(4-(*Tert*-Butyl)Phenyl)-2-Methyl-*N*-(Quinolin-8-yl)Propanamide the crude mixture was purified by gel permeation column chromatography (toluene) to afford the title compound in 85% yield as a pale white viscous liquid.

^{1}H NMR (500 MHz, CDCl3) δ 9.93 (s, 1H), 8.92 (dd, J = 7.7, 1.1 Hz, 1H), 8.60 (dd, J = 4.2, 1.6 Hz, 1H), 8.06 (dd, J = 8.3, 1.6 Hz, 1H), 7.57 (t, J = 8.0 Hz, 1H), 7.47 (dd, J = 8.3, 1.1 Hz, 1H), 7.32 (dd, J = 8.2, 4.2 Hz, 1H), 7.25 (d, J = 7.1 Hz, 2H), 7.22–7.15 (m, 6H), 7.15–7.10 (m, 1H), 3.54 (d, J = 13.2 Hz, 1H), 3.47 (d, J = 13.3 Hz, 1H), 2.81 (dd, J = 13.3, 4.4 Hz, 2H), 1.35 (s, 3H), 1.20 (s, 9H).

^{13}C NMR (126 MHz, CDCl3) δ 174.60, 149.01, 147.89, 138.57, 137.72, 135.95, 134.38, 134.22, 130.31, 129.96, 127.92, 127.68, 127.27, 126.28, 124.80, 121.29, 121.26, 116.17, 49.93, 46.38, 46.17, 34.17, 31.19, 19.66.

GC MS (EI) *m/z* (relative intensity): 436 (M+, 3), 345 (85), 289 (22), 201 (36), 171 (99), 145 (100), 117 (36), 91 (66).

2-Benzyl-3-(4-Fluorophenyl)-2-Methyl-*N*-(Quinolin-8-yl)Propanamide the crude mixture was purified by gel permeation column chromatography (toluene) to afford the titled compound in 56% yield as a pale white viscous liquid.

^{1}H NMR (500 MHz, CDCl3) δ 9.89 (s, 1H), 8.85 (dd, *J* = 7.6, 1.2 Hz, 1H), 8.61 (dd, *J* = 4.2, 1.6 Hz, 1H), 8.09 (dd, *J* = 8.2, 1.6 Hz, 1H), 7.55 (t, *J* = 7.9 Hz, 1H), 7.48 (dd, *J* = 8.3, 1.3 Hz, 1H), 7.36 (dd, *J* = 8.2, 4.2 Hz, 1H), 7.23–7.19 (m, 2H), 7.19–7.09 (m, 5H), 6.83(t, *J* = 8.8 Hz, 2H), 3.47 (dd, *J* = 13.3, 3.3 Hz, 2H), 2.75 (dd, *J* = 33.7, 13.3 Hz, 2H), 1.30 (s, 3H).

^{13}C NMR (126 MHz, CDCl3) δ 174.31, 161.64 (d, *J*F = 244.2 Hz), 148.08, 138.64, 137.47, 136.08, 134.08, 133.28 (d, *J*F = 3.2 Hz), 131.66 (d, *J*F = 7.8 Hz), 130.33, 128.02, 127.78, 127.29, 126.42, 121.50, 121.44, 116.25, 114.76 (d, *J*F = 21.0 Hz), 49.94 (d, *J*F = 1.1 Hz), 46.62, 45.69, 19.54.

GC MS (EI) *m/z* (relative intensity): 398 (M+, 3), 307 (34), 289 (34), 171 (100), 144 (37), 163 (27), 109 (55), 91 (51).

10.4.4   Deuterium-Labeling and KIE Experiments

Preparation of 1-Methyl-*d3*-*N*-(Quinolin-8-yl)Cyclohexanecarboxamide
A solution of LDA (30 mmol) in THF was prepared from diisopropylamine (32 mmol) and *n*-BuLi in hexane (1.6 M, 19 mL). To the LDA solution, ethyl cyclohexanecarboxylate (30 mmol) was added dropwise at −78 °C and the mixture was stirred at −78 °C for 1 h. Iodomethane-*d3* (40 mmol) was then added to the solution. The mixture was warmed to r.t. and stirred for 10 h before quenching at 0 °C with water. The mixture was extracted with Et$_2$O, and the combined organic layers were washed with brine. The organic layer was dried over MgSO$_4$, then evaporated in vacuo to give the crude ester.

To the ester was added a solution of sodium hydroxide (20 mL, 2 M) and methanol (30 mL). The mixture was stirred at 60 °C overnight. After removal of the methanol in vacuo, the pH of the mixture was adjusted to 2 with HCl (3 M). The product was extracted with Et$_2$O. The aqueous layer was saturated with NaCl and extracted again with Et$_2$O. The combined organic layers were washed with brine, dried over MgSO$_4$ and concentrated under vacuum to give the crude carboxylic acid, which was purified by silica gel chromatography (hexane: ethyl acetate = 5:1). The amide was prepared according to the following procedure: in a 100-mL flask was placed a carboxylic acid (11 mmol). SOCl$_2$ (10 mL) was slowly added to the solution. The reaction mixture was refluxed for 3 h at 85 °C, then the excess $SOCl_2$ was removed under vacuo to give the crude acid chloride. The crude acid chloride was diluted with dry dichloromethane (20 mL). A solution of 8-aminoquinoline (10 mmol) and NEt_3 (11 mmol) in dichloromethane (20 mL) was added dropwise to the acid chloride solution at 0 °C. The resulting mixture was allowed warm to r.t., and then stirred overnight. The mixture was quenched with saturated $NaHCO_3$ solution and extracted with CH_2Cl_2 three times. These extracts were combined and dried over $MgSO_4$. After evaporation under vacuo, the crude amide product was purified by silica gel chromatography (hexane: ethyl acetate = 10:1). (Reference: J. Am. Chem. Soc. **2009**, 131, 6898–6899.)

1-Methyl-*d*3-*N*-(Quinolin-8-yl)Cyclohexanecarboxamide the amide was purified by column chromatography on silica gel (*n*-hexane:EtOAc = 10:1) to afford the title compound as a yellow liquid. (Reference: J. Am. Chem. Soc. **1981**, 103, 436.)

^{1}H NMR (500 MHz, CDCl3) δ 10.28 (s, 1H), 8.82(dd, J = 7.6, 1.2 Hz, 1H), 8.78 (dd, J = 4.2, 1.7 Hz, 1H), 8.08 (dd, J = 8.3, 1.6 Hz, 1H), 7.50 (t, J = 7.9 Hz, 1H), 7.43 (dd, J = 8.2, 1.2 Hz, 1H), 7.38 (dd, J = 8.2, 4.2 Hz, 1H), 2.24–2.13 (m, 2H), 1.69–1.45 (m, 7H), 1.44–1.32 (m, 1H).

^{13}C NMR (126 MHz, CDCl3) δ 176.46, 148.06, 138.64, 136.11, 134.67, 127.77, 127.27, 121.36, 120.97, 116.03, 44.00, 35.56, 25.72, 22.84.

GC MS (EI) *m/z* (relative intensity): 271 (M+, 9), 171 (100), 144 (53), 116 (10).

Deuterium Labeling Experiment (Scheme 10.4)
A freshly prepared solution of (4-methoxyphenyl)magnesium bromide in THF (7.0 equiv, 3.5 mmol, 0.51 M) was slowly added using a gastight syringe to a mixture of 1-methyl-*d*3-*N*-(quinolin-8-yl)cyclohexanecarboxamide (1.0 equiv, 0.50 mmol) and $ZnBr_2 \cdot$ TMEDA (3.0 equiv, 1.5 mmol, 0.51 g) under argon. The resulted mixture was stirred at r.t. for 30 min, then a solution of Fe(acac)$_3$ (10 mol%, 0.05 mmol, 17 mg) and dppbz (10 mol%, 0.05 mmol, 23 mg) in anhydrous THF (2.0 mL) was added. The resulted black mixture was stirred at r.t. for 10 min. Then 1,2-dichloroisobutane (1.0 mmol, 115 μL) was added via syringe and the mixture was stirred at 50 °C for 22 h. After cooling to r.t., aqueous Rochelle's salt was added (5 mL), and the organic layer was extracted with diethyl ether (15 mL × 3). The combined organic layer was analyzed via GC and GC-MS, and then washed with brine, filtered through a pad of Florisil, dried over $MgSO_4$, and concentrated under reduced pressure. The crude mixture was analyzed by NMR using 1,1,2,2-tetrachloroethane as an internal standard and then purified by column chromatography on silica gel (hexane:ethyl acetate = 20:1–10:1) to afford the corresponding arylated product and the recovered starting material. The deuterium incorporation in the isolated product and recovered starting material was estimated

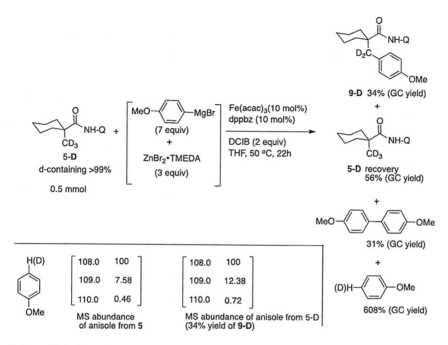

Scheme 10.4 Deuterium labeling experiment

by ¹H NMR. Deuterium incorporation into the recovered anisole was observed by GC MS.

1-(2D-(4-methoxyphenyl)methyl)-*N*-(quinolin-8-yl)cyclohexanecarboxamide: the amide was purified by column chromatography on silica gel (*n*-hexane: EtOAc = 10:1) to afford the titled compound as a white liquid.

¹H NMR (500 MHz, CDCl3) δ 10.01 (s, 1H), 8.81 (dd, *J* = 7.6, 1.3 Hz, 1H), 8.69 (dd, *J* = 4.2, 1.7 Hz, 1H), 8.12 (dd, *J* = 8.3, 1.6 Hz, 1H), 7.53 (t, *J* = 7.9 Hz, 1H), 7.47 (dd, *J* = 8.3, 1.3 Hz, 1H), 7.40 (dd, *J* = 8.2, 4.2 Hz, 1H), 7.01(d, *J* = 8.5 Hz, 2H), 6.61(d, *J* = 9.0 Hz, 2H), 3.61 (s, 3H), 2.27–2.18 (m, 2H), 1.78–1.67 (m, 2H), 1.63–1.45 (m, 5H), 1.38–1.27 (m, 1H).

¹³C NMR (126 MHz, CDCl3) δ 174.70, 158.08, 148.02, 138.70, 136.05, 134.46, 130.94, 129.04, 127.80, 127.37, 121.36, 121.07, 116.18, 113.19, 54.97, 49.44, 34.05, 25.91, 23.09.

GC MS (EI) *m/z* (relative intensity): 376 (M+, 16), 171 (57), 145 (64), 123 (100), 100 (14), 79 (10), 73 (10).

Intermolecular KIE experiment (Scheme 10.5) A freshly prepared solution of (4-methoxyphenyl)magnesium bromide in THF (7.0 equiv, 3.5 mmol, 0.51 M) was slowly added using a gastight syringe to a mixture of 1-methyl-$d3$-N-(quinolin-8-yl)cyclohexanecarboxamide (0.5 equiv, 0.25 mmol, 68 mg), 1-methyl-N-(quinolin-8-yl)cyclohexanecarboxamide (0.5 equiv, 0.25 mmol, 67 mg) and $ZnBr_2 \cdot$ TMEDA (3.0 equiv, 1.5 mmol, 0.51 g) under argon. The resulted mixture was stirred at r.t. for 30 min, then a solution of $Fe(acac)_3$ (10 mol%, 0.05 mmol, 17 mg) and dppbz (10 mol%, 0.05 mmol, 23 mg) in anhydrous THF (2.0 mL) was added. The resulted black mixture was stirred at r.t. for 10 min, then 1,2-dichloroisobutane (1.0 mmol, 115 μL) was added via syringe. The mixture was stirred at 50 °C for 1 h and then it was quenched by adding aqueous Rochelle's salt (5 mL). The organic layer was extracted with diethyl ether (15 mL × 3). The combined organic layer was analyzed via GC and GC-MS, and then washed with brine, passed through a pad of Florisil, dried over $MgSO_4$, and concentrated under reduced pressure. The crude mixture was analyzed by NMR using 1,1,2,2-tetrachloroethane as an internal standard and then purified by column chromatography on silica gel (hexane:ethyl acetate = 20:1–10:1) to afford the corresponding arylated products. The isolated products were further analyzed by ^1H NMR to determine the ratio of **9** and **9-D**.

^1H NMR (500 MHz, CDCl3) δ **10.00 (s, 1H)**, 8.80 (dd, J = 7.6, 1.0 Hz, 1H), 8.69 (dd, J = 4.2, 1.6 Hz, 1H), 8.12 (dd, J = 8.2, 1.5 Hz, 1H), 7.53 (t, J = 7.9 Hz, 1H), 7.47 (dd, J = 8.2, 1.1 Hz, 1H), 7.40 (dd, J = 8.2, 4.2 Hz, 1H), 7.00 (d, J = 8.6 Hz, 2H), 6.61 (d, J = 8.6 Hz, 2H), 3.61 (s, 3H), **2.90 (s, 1.60H)**, 2.33–2.18 (m, 2H), 1.83–1.66 (m, 2H), 1.62–1.45 (m, 5H), 1.38–1.25 (m, 1H).

KIE for Two Parallel Reactions (Scheme 10.6)
A freshly prepared solution of (4-methoxyphenyl)magnesium bromide in THF (7.0 equiv, 3.5 mmol, 0.51 M) was slowly added using a gastight syringe to a mixture of 1-methyl-$d3$-N-(quinolin-8-yl)cyclohexanecarboxamide (1.0 equiv, 0.50 mmol)

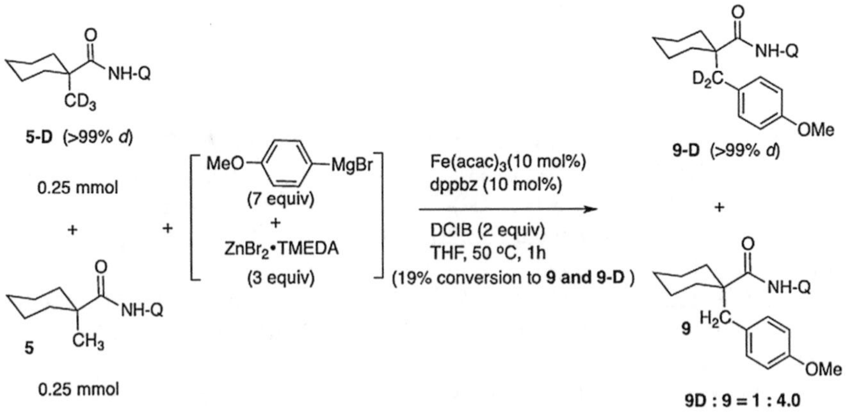

Scheme 10.5 Intermolecular KIE experiment

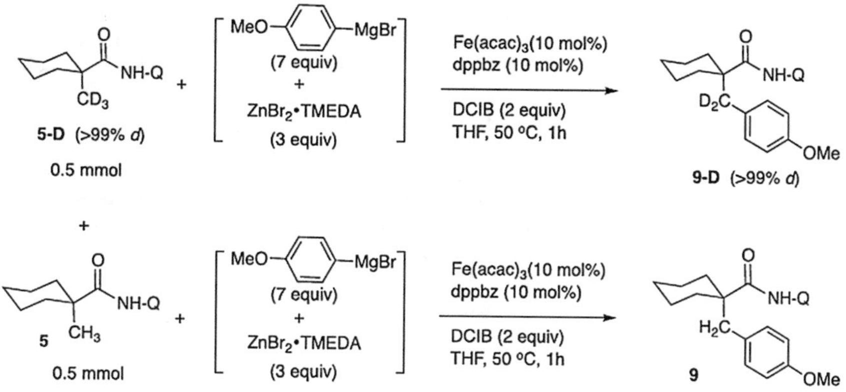

Scheme 10.6 KIE for two parallel reactions

and $ZnBr_2 \cdot TMEDA$ (3.0 equiv, 1.5 mmol, 0.51 g) under argon. Another freshly prepared solution of (4-methoxyphenyl)magnesium bromide in THF (7.0 equiv, 3.5 mmol, 0.51 M) was slowly added using a gastight syringe to a mixture of 1-methyl-*N*-(quinolin-8-yl)cyclohexanecarboxamide (1.0 equiv, 0.5 mmol), and $ZnBr_2 \cdot TMEDA$ (3.0 equiv, 1.5 mmol, 0.51 g) under argon. Under otherwise identical reaction conditions, the two reaction mixtures were stirred at r.t. for 30 min, then a solution of $Fe(acac)_3$ (10 mol%, 0.05 mmol, 17 mg) and dppbz (10 mol%, 0.05 mmol, 23 mg) in 2.0 mL anhydrous THF were added to each reaction. The resulted black mixtures were stirred at r.t. for 10 min, then 1,2-dichloroisobutane (1.0 mmol, 115 μL) and tridecane (0.5 mmol, 125 μL, as an internal standard) were added via syringe to each reaction. The mixtures were stirred at 50 ± 1.5 °C. The reactions were sampled every 10 min and analyzed by GC. The GC yield was calculated after calibrating the response of GC. The average ratio of the two target products (**9** and **9-D**) for the first 60 min reaction is reported as the KIE value.

References

1. (a) Bolm, C. (2009). *Nature. Chemistry, 1,* 420. (b) Nakamura, E., & Sato, K. (2011). *Nature Material, 10,* 158–161.
2. Selected reviews: (a) Bolm, C., Legros, J., Le Paih, J., & Zani, L. (2004). *Chemical Reviews, 104,* 6217–6254. (b) Plietker, B. Ed. (2008). *Iron Catalysis in Organic Chemistry* Wiley-VCH: Weinheim, Germany. (c) Enthaler, S., Junge, K., & Beller, M. (2008). *Angewandte Chemie International Edition, 47,* 3317–3321. (d) Sherry, B. D., & Fürstner, A. (2008). *Accounts of chemical research, 41,* 1500–1511. (e) Czaplik, W. M., Mayer, M., Cvengros, J. & Jacobi von Wangelin, A. (2009). *Chemistry and Sustainable Chemistry, 2,* 396–417.
3. Dyker, G. Ed. *Handbook of C–H Transformations* Wiley-VCH: Weinheim, Germany, (2005).
4. Sun, C.-L., Li, B.-J., & Shi, Z.-J. (2011). *Chemical Reviews, 111,* 1293–1314.

5. (a) Godula, K., & Sames, D. (2006). *Science, 312*, 67–72. (b) Chen, X., Engle, K. M., Wang, D.-H., & Yu, J.-Q. (2009). *Angewandte Chemie International Edition, 48*, 5094–5115. (c) Jazzar, R., Hitce, J., Renaudat, A., Sofack-Kreutzer, J., & Baudoin, O. (2010). *Chemistry-A European Journal, 16*, 2654–2672. (d) Baudoin, O. (2011). *Chemical Society Reviews, 40*, 4902–4911. (e) Li, H., Li, B.-J., & Shi, Z.-J. (2011). *Catalysis Science & Technology, 1*, 191–206.

6. Selected recent examples: (a) Ano, Y., Tobisu, M., & Chatani, N. (2011). *Journal of the American Chemical Society, 133*, 12984–12986. (b) Wasa, M., Chan, K. S. L., Zhang, X.-G., He, J., Miura, M., & Yu, J.-Q. (2012). *Journal of the American Chemical Society, 134*, 18570–18572. (c) Rousseaux, S., Liégault, B., & Fagnou, K. (2012). *Chemical Science, 3*, 244–248. (d) Zhang, S.-Y., He, G., Nack, W. A., Zhao, Y., Li, Q., & Chen, G. (2013). *Journal of the American Chemical Society, 135*, 2124–2127. (e) Shang, Y., Jie, X., Zhou, J., Hu, P., Huang, S., & Su, W. (2013). *Angewandte Chemie International Edition, 52*, 1299–1303.

7. (a) Li, Z., Cao, L., & Li, C.-J. (2007). *Angewandte Chemie International Edition, 46*, 6505–6507. (b) Zhang, Y., & Li, C.-J. (2007). *European Journal of Organic Chemistry*, 4654–4657. (c) Li, Z., Yu, R., & Li, H. (2008). *Angewandte Chemie International Edition, 47*, 7497–7500. (d) Li, Y.-Z., Li, B.-J., Lu, X.-Y., Lin, S., & Shi, Z.-J. (2009). *Angewandte Chemie International Edition, 48*, 3817–3820. (e) Volla, C. M. R., & Vogel, P. (2009). *Organic Letters, 11*, 1701–1704. (f) Singh, P. P., Gudup, S., Ambala, S., Singh, U., Dadhwal, S., Singh, B., Sawant, S. D., & Vishwakarma, R. A. (2011). *Chemical Communications, 47*, 5852–5854.

8. (a) Yoshikai, N., Mieczkowski, A., Matsumoto, A., Ilies, L., & Nakamura, E. (2010). *Journal of the American Chemical Society, 132*, 5568–5569. (b) Sekine, M., Ilies, L., & Nakamura, E. (2013). *Organic Letters, 15*, 714–717.

9. (a) Nakamura, E., & Kuwajima, I. (1977). *Journal of the American Chemical Society, 99*, 7360–7361. (b) Nakamura, E., & Kuwajima, I. (1984). *Journal of the American Chemical Society, 106*, 3368–3370. (c) Aoki, S. Fujimura, T. Nakamura, E., & Kuwajima, I. (1988). *Journal of the American Chemical Society, 110*, 3296–3298.

10. (a) Zaitsev, V. G., Shabashov, D., & Daugulis, O. (2005). *Journal of the American Chemical Society, 127*, 13154–13155. (b) Shabashov, D., & Daugulis, O. (2010). *Journal of the American Chemical Society, 132*, 3965–3972.

11. Li, B.-J., & Shi, Z.-J. (2012). *Chemical Society Reviews, 41*, 5588–5598.

12. Nakamura, E., & Yoshikai, N. (2010). *The Journal of Organic Chemistry, 75*, 6061–6067.

13. (a) Norinder, J., Matsumoto, A., Yoshikai, N., & Nakamura, E. (2008). *Journal of the American Chemical Society, 130*, 5858–5859. (b) Yoshikai, N., Matsumoto, A., Norinder, J., & Nakamura, E. (2009). *Angewandte Chemie International Edition, 48*, 2925–2928. (c) Ilies, L., Asako, S., & Nakamura, E. (2011). *Journal of the American Chemical Society, 133*, 7672–7675. (d) Matsumoto, A., Ilies, L., & Nakamura, E. (2011). *Journal of the American Chemical Society, 133*, 6557–6559.

14. (a) Cahiez, G., Chaboche, C., Mahuteau-Betzer, F., & Ahr, M., (2005).*Organic Letters, 7*, 1943–1946. (b) Nagano, T., & Hayashi, T. (2005). *Organic Letters, 7*, 491–493.

15. Nakamura, Y., Ilies, L., & Nakamura, E. (2011). *Organic Letters, 13*, 5998–6001.

16. Yoshikai, N., Asako, S., Yamakawa, T., Ilies, L., & Nakamura, E. (2011). *Chemistry – An Asian Journal, 6*, 3059–3065.

17. (a) Baker, M. V., & Field, L. D. (1987). *Journal of the American Chemical Society, 109*, 2825–2826. (b) Field, L. D., & Baker, M. V. (1999). *Australian Journal of Chemistry, 52*, 1005–1011. (c) Ohki, Y., Hatanaka, T., & Tatsumi, K. (2008). *Journal of the American Chemical Society, 130*, 17174–17186. (d) Xu, G. Q., & Sun, H., Li, X. (2009). *Organometallics, 28*, 6090–6095.

Chapter 11
Iron-Catalyzed Directed C(sp^2)–H Bond Functionalization with Organoboron Compounds

Abstract Iron, the most abundant transition metal on earth, has so far not received enough attention in catalytic organic synthesis because the catalytic activity of organoiron species is often difficult to control, resulting in narrow applicability. We report here that an iron-catalyzed C–H functionalization reaction allows the coupling of a wide variety of combinations of aromatic, heteroaromatic and olefinic substrates, and alkenyl, alkyl and aryl boron compounds among the reactions catalyzed by precious metals such as palladium. We rationalize these results by the involvement of an organoiron(III) reactive intermediate that is responsible for the C–H bond activation process, and a metal-to-ligand charge transfer that enables smooth catalytic turnover via an iron(I) intermediate.

11.1 Introduction

Efficiency and sustainability are two requirements in modern synthetic organic chemistry, and therefore functionalization of an intrinsically unreactive C–H bond has received wide attention for its synthetic efficiency and the production of potentially less waste than the classical substitution reactions [1]. C–H functionalization catalyzed by palladium [2–7] and other precious metals [8–11] has found widespread use for coupling of a hydrocarbon bearing a directing group and an organoboron compound, which is shelf-stable, readily available and tolerant of a variety of functional groups. These reactions, however, are not without limitations;

© Springer Nature Singapore Pte Ltd. 2017   197
R. Shang, *New Carbon–Carbon Coupling Reactions Based on Decarboxylation and Iron-Catalyzed C–H Activation*, Springer Theses,
DOI 10.1007/978-981-10-3193-9_11

for instance, they are unsuitable for alkenyl–alkenyl and alkenyl–arene couplings (Fig. 11.1a) where the diene and styrene products are prone to isomerize [8] or to undergo further transformations [4]. One probable reason for this limitation is the preferred interaction of these metals with the π-bonds in the product rather than with the C(sp^2)–H σ-bond in the starting material. Hence, we conjectured that a transition metal with a minimum π-back donation ability prefers C–H bond activation via σ-bond metathesis [12] over interaction with the π-bonds. We report here that iron catalysis [13–17] complements or surpasses precious metal catalysis for

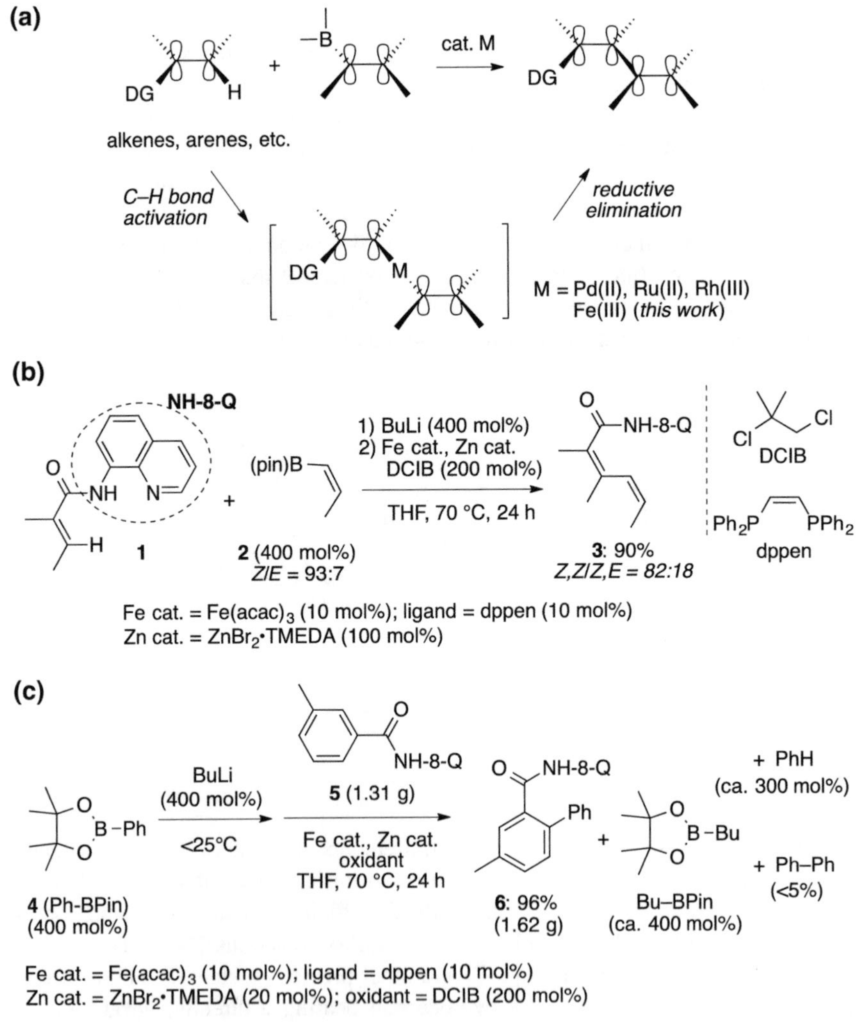

Fig. 11.1 C(sp^2)–H functionalization with organoboronate ester under iron–zinc cooperative catalysis. **a** General scheme for C(sp^2)–H functionalization with organoboron compound. **b** Stereospecific diene synthesis. **c** Procedure for *ortho*-functionalization of aromatic amide

substrate scope and applicability. Iron also benefits from advantages over precious metals, such as abundance in nature, low cost, and lack of toxicity [18].

Despite the prevalent view that a low-valent iron complex is required for C–H bond activation [19–21], we found that an organoiron(III) intermediate effectively performs stereo- and chemoselective $C(sp^2)$–H functionalization when it is generated properly without contamination with low-valent iron species. The reaction reported here utilizes catalytic amounts of $Fe(acac)_3$ and a zinc halide [22], a bidentate diphosphine ligand ((Z)-1,2-bis(diphenylphosphino)ethene: dppen), an alkene or arene substrate possessing a directing group, an organic pinacol boronate (R–Bpin) and 1,2-dichloroisobutane (DCIB) that keeps the reaction conditions mildly oxidative (Fig. 11.1b and c). R–Bpin is activated in situ with a stoichiometric amount of butyllithium (BuLi) [23] and an essentially catalytic amount of zinc halide for the transfer of the R group to the substrate (Fig. 11.1c). This new preparation of organozinc species from organoboronates allows selective formation of organoiron(III) intermediates, and hence is much superior to the use of organomagnesium or zinc reagents that cause uncontrollable reduction of the iron (III) species and ensuing side reactions [24–26]. Thus, the iron–zinc cooperative catalysis cross couples efficiently a variety of aromatic, heteroaromatic and olefinic substrates with a variety of alkenyl, alkyl and aryl boron compounds, as highlighted by the synthesis of a (Z,Z)-diene **3** by coupling of an *E*-alkenylamide **1** and *Z*-boronate ester **2** (Fig. 11.1b, c). The reaction requires the use of a one-electron-accepting ligand such as dppen. This experimental observation together with a computational study suggests that a metal-to-ligand charge transfer (MLCT) process facilitates the catalytic turnover of iron(I) species that forms at the end of the catalytic cycle.

11.2 Results and Discussion

11.2.1 Investigation of the Reaction Conditions

The experimental procedure optimized for the reaction of 3-methyl-*N*-(quinolin-8-yl) benzamide (**5**) and phenylboronic acid pinacolester (Ph–Bpin, **4**) (Fig. 11.1c) is the following: a solution of BuLi (slightly less than 20 mmol) in hexane was slowly added to **4** (20 mmol) in THF at −78 °C, the mixture was allowed to warm to room temperature and then a solution of $Fe(acac)_3$ (0.5 mmol), dppen (0.5 mmol), $ZnBr_2 \cdot$ TMEDA (1.0 mmol), and **5** (1.31 g, 5.0 mmol) in THF was added. DCIB (10 mmol) was added and the reaction mixture was stirred at 70 °C for 24 h to obtain an *ortho*-phenylated amide **6** in 96% yield (1.62 g) after purification by column chromatography on silica gel. Careful exclusion of moisture was needed to ensure reproducibility. The overall stoichiometry of the reaction is as follows (Fig. 11.1c): the 4 equiv of the butyl group originating from BuLi stays quantitatively on the boron atom to liberate 4 equiv of an anionic phenyl group, 2 equiv are consumed for removal of two hydrogen atoms from the substrate, and 1 equiv for delivery of a phenyl group. We can save 1 equiv of Ph–Bpin/BuLi by prior removal of the amide proton with a Grignard reagent at the expense of ca. 30% decrease of the product yield. We noted that for both catalytic and stoichiometric reactions (vide infra), the fourth equivalent is necessary for complete conversion. The benefit of the use of the quinolylamide was shown by an experiment in which we used a nonchelating *N*-methylbenzamide and obtained the *ortho*-phenylated product in 40% yield and biphenyl in 52% yield.

The study on the reaction parameters includes the variation of iron salt, zinc(II) salt, oxidant, and ligand (Fig. 11.2a). The most notable is the success of the reaction using stoichiometric iron(III) without additional oxidant (entry 1). We thus utilized 1 equiv of the quinolylamide 5, 4 equiv of Ph–Bpin 4/BuLi, and 1 equiv each of Fe (acac)$_3$, dppen and ZnBr$_2$ · TMEDA without any oxidant, and obtained the desired product 6 in 95% yield together with biphenyl in 12% yield and Bu–Bpin in quantitative yield. At the end of the reaction, iron was recovered as black precipitates. When we decreased the amount of Ph–Bpin 4 to 3 equiv, the yield decreased to 58%, and 36% of the amide 5 was recovered (entry 2). The necessary use of excess phenyl donor suggests the formation of an iron(III) species bearing more than one phenyl group. The reaction without dppen and TMEDA gave back 5 and produced biphenyl (44%) (entry 3). Therefore, dppen is crucial for C–H activation and suppression of the decomposition of the diphenyliron(III) intermediate. The use of Fe(II) instead of Fe(III) afforded 6 in only 38% yield with 56% recovery of 5 (entry 4). Taking these results together, we consider that a diphenyliron(III) species chelated to the anion of 5 (Fig. 11.2a) is a reactive intermediate of the C–H functionalization, which then forms an iron(I) species (D and E) as a product. Thus, we may need to revise our previous hypothesis that a reduced iron species cleaves the C–H bond [27, 28].

Under the catalytic conditions (10 mol% Fe(acac)$_3$), we need DCIB to recycle the reduced iron(I) species back to iron(III) reactive species. Thus, catalytic conditions using 10 mol% of Fe(acac)$_3$, 20 mol% of ZnBr$_2$ · TMEDA and excess DCIB gave 6 in 96% yield and a trace amount of biphenyl (entry 5). The use of a larger amount of a zinc salt entirely suppressed the biphenyl formation (entries 6–9), and we therefore used 100 mol% ZnBr$_2$ · TMEDA for the study of the reaction scope (Fig. 11.3). Without a zinc salt or without Fe(acac)$_3$, no reaction took place (entries 10 and 11). We found that a catalytic amount of Fe(acac)$_2$ fails entirely (entry 12). With this result, taken together with the poor yield in entry 4, we conclude that an organoiron(II) species is not a reactive species of C–H bond activation. The reaction without dppen ligand gave 6 in low yield (entry 13). Without DCIB, the catalytic turnover number was essentially one (entry 14). 1,2-Dichloroethane (DCE), which is typically used as a solvent in organic synthesis, also promotes the catalytic cycle, albeit in slightly lower yield (entry 15).

The effect of the ligand is described in Fig. 11.2b. Among the ligands investigated, dppen, dppBz, 1,10-Phen (1,10-phenanthroline), and dtbpy (4,4′-di-*tert*-butyl-2,2′-bipyridine) are much better than others, affording the desired product 6 in 85–99% yield.

(a)

entry	Fe (mol%)	Zn (mol%)	oxidant (mol%)	ligand (mol%)	6 (%)[a]	5 (%)[b]	Ph–Ph[b]
1	Fe(acac)$_3$ (100)	ZnBr$_2$·TMEDA (100)	none	dppen (100)	95	0	12
2[c]	Fe(acac)$_3$ (100)	ZnBr$_2$·TMEDA (100)	none	dppen (100)	58[b]	36	13
3	Fe(acac)$_3$ (100)	ZnBr$_2$ (100)	none	none	0	91[d]	44
4[c]	Fe(acac)$_2$ (100)	ZnBr$_2$·TMEDA (100)	none	dppen (100)	38[b]	56	7
5	Fe(acac)$_3$ (10)	ZnBr$_2$·TMEDA (20)	DCIB (200)	dppen (10)	96	0	<5
6	Fe(acac)$_3$ (10)	ZnBr$_2$·TMEDA (50)	DCIB (200)	dppen (10)	99	0	0
7	Fe(acac)$_3$ (10)	ZnBr$_2$·TMEDA (100)	DCIB (200)	dppen (10)	99	0	0
8	Fe(acac)$_3$ (10)	ZnCl$_2$·TMEDA (50)	DCIB (200)	dppen (10)	99	0	0
9	Fe(acac)$_3$ (10)	ZnBr$_2$ (50)	DCIB (200)	dppen(10)	95	0	0
10	Fe(acac)$_3$ (10)	none	DCIB (200)	dppen (10)	0	96[a]	0
11	none	ZnBr$_2$·TMEDA (100)	DCIB (200)	dppen (10)	0	98[a]	0
12	Fe(acac)$_2$ (10)	ZnBr$_2$·TMEDA (50)	DCIB (200)	dppen (10)	0	64	<5
13	Fe(acac)$_3$ (10)	ZnBr$_2$·TMEDA (100)	DCIB (200)	none	27	72[a]	6
14	Fe(acac)$_3$ (10)	ZnBr$_2$·TMEDA (100)	none	dppen (10)	13	84[a]	0
15	Fe(acac)$_3$ (10)	ZnBr$_2$·TMEDA (100)	DCE[e] (200)	dppen (10)	87	11	0

[a] Yield determined by NMR using 1,1,2,2-tetrachloroethane as an internal standard. [b] Yield determined by GC using tridecane as an internal standard. [c] With 300 mol% Ph–Bpin/BuLi; ca 300 mol% Bu–Bpin was generated (GC). [d] Upon quenching with D$_2$O, no D was incorporated into the *ortho* site of **5** (GC-MS). [e] DCE = 1,2-dichloroethane.

(b)

Ph$_2$P⁔PPh$_2$
dppen
99%

Ar$_2$P⁔PAr$_2$
dppBz (Ar = Ph): 96%
TMS-SciOPP (Ar = 3,5-bis(trimethylsilyl)phenyl): 47%

Ph$_2$P⁔PPh$_2$
dppe
8%

PPh$_2$ / PPh$_2$ (XantPhos)
XantPhos
26%

Fe / PPh$_2$ / PPh$_2$
dppf
<5%

1,10-Phen
85%

dtbpy
93%

Reaction conditions: Fe(acac)$_3$/**ligand** (10 mol%), ZnBr$_2$·TMEDA (50 mol%) and DCIB (200 mol%)

Fig. 11.2 a Effect of several key parameters on the stoichiometric and catalytic *ortho*-phenylation of **5** with Ph–Bpin/BuLi. **b** Effect of ligand on the yield of product **6**

11.2.2 Exploration of the Substrate Scope

The procedure optimized for the phenylation of amide **5** is applicable with little modification to the coupling of alkenyl, aryl and heteroaryl compounds bearing the quinolylamide group, and alkenyl, alkyl and aryl Bpin derivatives (Fig. 11.3). Thus, the iron catalysis rivals the hydrocarbon–boronate coupling using precious metals for the formation of aryl–aryl and aryl–alkyl bonds, and it is a unique

1) BuLi (400 mol%)
2) Fe(acac)₃/dppen (10 mol%)
   ZnBr₂·TMEDA (100 mol%)

DCIB (2 equiv)
THF, 70 °C, 16–24 h

R–B (pinacol ester) (400 mol%)

(a)

7: 88% yield
>99% *Z,E*

8: 71% yield
>99% *Z,E*

9: 75% yield
>99% *Z,E*

10: 87% yield
>99% *Z*

3: 90% yield
Z,Z/Z,E = 82:18[a]

11: 79% yield
>99% *Z,E*

12: 66% yield
>99% *Z,E*

13: 91% yield
>99% *Z,E*

14: 65% yield
>99% *Z,E,E*

15: 81% yield
>99% *Z,E,E*

16: 89%
>99% *Z,E*

17: 64%

18: 62%[b]
(*E/Z* = 73: 27)

19: 86%
E/Z >99:1

20: 88%

21: 68%

22: 86%
E/Z = 16:84[a]

23: 84%
E/Z >99:1

24: 51%
E/Z >99:1

(b)

R = H   mono (**25a**) 77%
         di   (**25b**) 19%
R = Me   mono (**26a**) 55%
         di   (**26b**) 38%

27: 83%

28: 94%

R = Me   (**6**) 96%
    OMe   (**29**) 92%
    CF₃   (**30**) 89%
    CN   (**31**) 56%
    CO₂Me (**32**) 74%

X = F   (**33**) 95%
    Cl   (**34**) 96%
    Br   (**35**) 76%
    OMe   (**36**) 97%
    SMe   (**37**) 94%
    NMe₂ (**38**) 92%

39: 51%[c]

40: 96%

41: 95%

42: 90%

43: 88%

44: 83%[c]

45: 72%[c]

46: 52%[c]

(c)

47: 98%[c,d]

mono (**48a**) 43%
di   (**48b**) 54%[c,d]

49: 99%[c,d]

50: 72%[c,d]

◀ **Fig. 11.3** Reaction scope. **a** Alkenyl–alkenyl, alkenyl–aryl and aryl–alkenyl coupling using alkenyl Bpin compounds. **b** Aryl–aryl and aryl–alkyl coupling using aryl and alkyl Bpin compounds. **c** Substrates bearing a pyridine and pyrazole directing group. [a]Stereochemical purity of the starting organoboron was E:Z = 7:93. [b]At 50 °C. [c]Using sec-BuLi as a base. [d]At 30 °C

catalytic system that is effective for stereospecific aryl–alkenyl and alkenyl–alkenyl coupling for the synthesis of diene, triene and styrene derivatives. The iron(III) catalysis circumvented the major side reactions encountered in the precious metal-catalyzed C–H functionalization, such as positional isomerization [8] and cyclization of the initial olefin product [4].

Figure 11.3a illustrates the *syn*-selective C–H functionalization of olefins bearing a quinolylamide, which can be carried out with tolerance of diene, triene, ether, silylether, and chloride groups. The stereochemistry of E-boronates was entirely retained, while that of (Z)-1-propenylboronate **2** (E:Z = 7:93) was retained to an extent of 80–90%, as illustrated by the synthesis of (Z,Z)-2,3-dimethylhexa-2,4-dienoic acid amide **3** with 100% selectivity for the 2Z bond and 82% selectivity for the 3Z bond (88% retention). (Z,E,E)-Trienes **14** and **15** were similarly prepared in good yield with 100% retention of the E-geometry in the dienylboronate. Alkenylamides were also arylated with arylboronates (e.g., **16** and **17**). While a tiglic amide **1** gave the product **16** with 100% Z selectivity, an acrylic acid amide gave a cinnamic acid amide **18** with only 27% Z selectivity because of in situ isomerization of the initial Z-product. An electron-transfer mechanism of isomerization is suspected [26]. The bottom of Fig. 11.3a shows the synthesis of styrene derivatives, including the synthesis of (Z)-1-propenyl product **22** from **2** (84% Z from 93% Z-boronate) as well as a *trans*-stilbene product **19** (starting from 100% E styrene boronate) and a 2-propenylated product **20**. We did not observe any ring-opening side reactions in the synthesis of vinylcyclopropane product **23**.

Figure 11.3b summarizes the reactions of aryl and heteroaryl substrates with aryl and alkyl boronate esters. When unsubstituted or *para*-substituted benzeneamides were used as a substrate, mono- and diarylation occurred to give **5a,b** and **26a,b**, respectively. A *meta*-substituent in the benzeneamide shut off the arylation in the nearby *ortho*-position, resulting in the exclusive formation of monoarylated products **6**, **28–41**. *Ortho*-steric hindrance in the aryl boronate part was tolerated as illustrated for *ortho*-tolyl boronate, which gave **39** in 51% yield.

Functional group tolerance is an asset of the boron reagent over more reactive Grignard and zinc reagents, as highlighted in Fig. 11.3b. For instance, aryl fluoride (**33**), chloride (**34**), bromide (**35**), amine (**38**), nitrile (**31**) and ester (**32**) are tolerated. Interestingly, the reaction also tolerates a sulfi group (**37**)—a typical catalyst poison. Heteroatom-containing thiophene and indole also served as a good substrate to give amides **42** and **43** in ca. 90% yield, could introduce a butyl group to obtain **44** in 83% yield with little evidence competitive β-elimination, and also a cyclopropyl group to obtain **45** in 73% (Fig. 11.3b). The *ortho*-methoxybenzamide **46** may serve as a precursor for synthesis of lunularic acid natural products [29].

We can couple Ph–Bpin with arene and alkene substrates bearing a pyridine or a pyrazole directing group in good to excellent yield (Fig. 11.3c, compounds **47–50**). These substrates are, however, prone to give a larger amount of the biphenyl side product, suggesting that anionic chelation by the quinolylamide anion provides a uniquely effective environment for the iron(III) catalysis. The biphenyl formation was suppressed to a synthetically viable level by carrying out the reaction at 30 °C instead of the standard 70 °C.

11.2.3 Mechanistic Study

The BuLi-mediated process for the phenyl transfer from Ph–Bpin to zinc was probed first (Fig. 11.4a). ^{11}B NMR studies showed that addition of BuLi in hexane

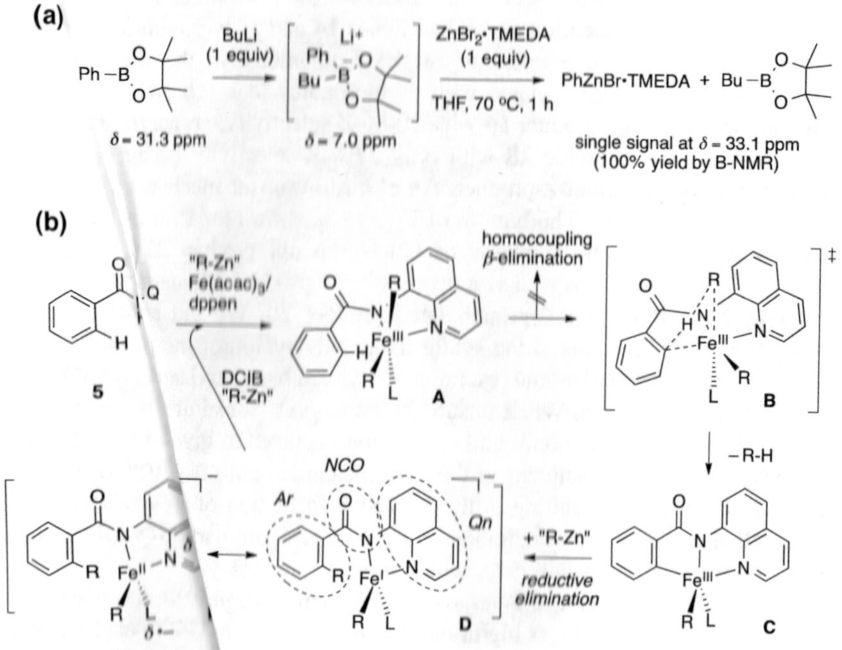

Fig. 11.4 Mechanistic information. **a** Boron–zinc exchange reaction. **b** A rationale for the iron (III)–iron(I) catalytic cycle. **c** density and charge in **D** for the sextet (L = dppen, R = Me) calculated at the B3LYP/SV(P) of theory

to Ph–Bpin in THF (δ = 31.3 ppm) resulted in quantitative formation of the corresponding lithium borate (δ = 7.0 ppm). Upon addition of a Zn(II) salt, the borate signal disappeared [30] and a single signal due to Bu–Bpin at δ = 33.1 ppm appeared, indicating quantitative transfer of the phenyl group to the zinc atom [31]. Because the 33.1-ppm chemical shift is the same as that of pure Bu–Bpin, we consider that there is little interaction between the Bu–Bpin and the phenyl zinc reagent.

We obtained several lines of information on the mechanism of the iron catalysis. First, the stoichiometric reactions discussed for Fig. 11.1c and the superiority of the quinolylamide over a monodentate donor (cf. Fig. 11.3c) indicate that a chelated organoiron(III) intermediate is responsible for the C–H functionalization. Second, the required use of excess R–Bpin/BuLi in both stoichiometric and catalytic reactions suggests that more than one organic group on the iron atom is required for either the C–H bond cleavage and/or the C–R bond formation. Third, a π-accepting ligand (e.g., dppen and 1,10-Phen) is necessary. Taken together, these observations suggest an active role of a chelated diorganoiron(III) species in the C–H functionalization and stabilization of an iron(I) intermediate by the ligand in the late stage of the catalytic cycle.

A mechanistic rationale is shown in Fig. 11.4b. A plausible first intermediate is a trigonal bipyramidal iron(III) species **A** that has one R group in an apical position and another in an equatorial position. The rigid bidentate quinolylamide anion [32, 33] probably prevents decomposition of this organoiron(III) intermediate [34] through homocoupling and β-elimination pathways. **A** then goes to a transition state **B**, where only one of the R groups can react with the C(sp^2)–H bond for stereoelectronic reasons. We expect that this R–Fe(III) species removes the *ortho*-hydrogen via σ-bond metathesis (i.e., proton removal without change of the oxidation state) to form a trichelate iron(III) **C** [35, 36]. We may consider that the excess R anion source in the reaction mixture produces an iron(I)ate complex **D** after reductive elimination so that it can readily react further with **5**, DCIB and a zinc reagent to regenerate **A**. We expect that MLCT with dppen creates a formal resonance involving iron(I) **D** and iron(II) species **E**. This conjecture was corroborated by DFT calculations at the B3LYP/SVP level of theory. A sextet state is more stable than doublet (+15.7 kcal/mol) and quartet (+8.0 kcal/mol) states (Fig. 11.4c), and it shows considerable spin density (0.85) and charge (−0.59) on the dppen ligand as well as on the quinolylamide moiety (0.23 and −0.20, respectively), making **D** behave as iron(II) **E** (the same was found for doublet and quartet; see SI for details). The stabilizing role of dppen would be much larger during the catalytic turnover where the quinolylamide ligand is detached from the iron(I) atom.

11.3   Conclusion

The iron-catalyzed C–H functionalization reported here allows us to synthesize the widest variety of C–H functionalization products among comparable methods catalyzed by precious metals. The versatility of the method is illustrated by the synthesis of dienes, trienes, and styrene derivatives with retention of stereochemistry in the starting materials, as enhanced by the commercial availability of a variety of organoboronate compounds because of the popularity of the Suzuki–Miyaura coupling reactions [37]. Experimental and theoretical studies revealed several key mechanistic features of the new iron catalysis. The new method for in situ generation of an organozinc reagent from R–Bpin effects selective formation of R–Fe(III), and together with the use of a bidentate quinolylamide directing group excludes the reduction of iron species that is so prone to occur under previously reported conditions. We obtained experimental evidence for a decisive role of iron (III) intermediates in the C–H bond activation via σ-bond metathesis, and also the MLCT stabilization of the iron(I) species by dppen, which will facilitate catalytic turnover [38]. The effectiveness of DCIB in oxidizing an iron(I) product back to an iron(III) catalyst is also responsible for the tolerance of sensitive olefin-rich products as well as a variety of functional groups. An example of product isomerization during the reaction of an acrylic amide suggests undesired electron transfer from iron(I) to the product, the suppression of which will be the subject of future studies. Overall, the reaction reported in this article represents a rare demonstration of the synthetic advantage of organoiron catalysis over catalysis with precious metals.

11.4   Experimental Section and Compound Data

11.4.1   General Information

All the reactions dealing with air- or moisture-sensitive compounds were carried out in a dry reaction vessel under a positive pressure of argon. Air-and moisture-sensitive liquids and solutions were transferred via syringe or Teflon cannula. Analytical thin-layer chromatography was performed using glass plates precoated with 0.25-mm 230–400 mesh silica gel impregnated with a fluorescent indicator (254 nm). Thin-layer chromatography plates were visualized by exposure to ultraviolet light (UV). Organic solutions were concentrated by rotary evaporation at ca. 15 Torr (evacuated with a diaphragm pump). Flash column chromatography was performed as described by Still et al. (J. Org. Chem. **1978**, 43, 2923–2924), employing Kanto Silica gel 60 (spherical, neutral, 140–325 mesh). Gas–liquid chromatographic (GLC) analysis was performed on a Shimadzu 14A or 14B machine equipped with glass capillary column HR-1 (0.25-mm i.d. × 25 m). Gel permeation column chromatography was performed on a Japan Analytical Industry LC-908 (eluent:toluene) with JAIGEL 1H and 2H polystyrene columns.

Proton nuclear magnetic resonance (^{1}H NMR) and carbon nuclear magnetic resonance (^{13}C NMR) spectra were recorded with JEOL ECA-500 NMR or JEOL ECX-400 spectrometers. Chemical data for protons are reported in parts per million (ppm, δ scale) downfield from tetramethylsilane and are referenced internally to tetramethylsilane. Carbon nuclear magnetic resonance spectra (^{13}C NMR) were recorded at 125 or 100 MHz. Chemical data for carbons are reported in parts per million (ppm, δ scale) and are referenced to the carbon resonance of the solvent (CDCl$_3$: = 77.0). The data is presented as follows: chemical shift, multiplicity (s = singlet, d = doublet, t = triplet, q = quartet, m = multiplet and/or multiplet resonances, br = broad), coupling constant in Hertz (Hz), and integration. Mass spectra (GS MS) are taken at SHIMADZU Parvum 2 gas chromatograph mass spectrometer.

Commercial reagents were purchased from Tokyo Kasei Co., Aldrich Inc., and other commercial suppliers and were used after appropriate purification. Anhydrous ethereal solvents (stabilizer-free) were purchased from WAKO Pure Chemical and purified by a solvent purification system (GlassContour) equipped with columns of activated alumina and supported copper catalyst (Q-5) prior to use (Organometallics **1996**, 15, 1518–1520). The content of water in the solvents or the substrates was confirmed to be less than 30 ppm by Karl–Fischer moisture titrator. The boronate reagents were directly purchased from Aldrich Inc. or prepared from the corresponding boronic acids using the method reported in the literature (J. Am. Chem. Soc. **2013**, 135, 2552–2559), and were dried over molecular sieves (4 Å) prior to use. Fe(acac)$_3$ (99.9+%) was purchased from Aldrich Inc. and used as received. Iron salts from other sources are mentioned in the text where appropriate. (Z)-1,2-bis (diphenylphosphino)ethene (dppEn) was purchased from Aldrich Inc. and used as received. 1,2-Dichloroisobutane (DCIB) was purchased from Tokyo Kasei Co. and dried over molecular sieves prior to use. ZnBr$_2$ · TMEDA and ZnCl$_2$ · TMEDA were prepared according to the literature procedure (Chem. Lett. **1977**, 679–682).

Caution: we observed that contamination of the reaction mixture by atmospheric moisture significantly reduces the yield. We sometimes observed complete suppression of the reaction when using incompletely anhydrous conditions, and we ascribe this to the poisoning of the iron catalyst by hydroxide species. For best results, all the experiments described herein must be performed under rigorous anhydrous and air free conditions.

11.4.2 Experimental Procedures

11.4.2.1 Investigation of the Key Reaction Parameters

Scheme 11.1 shows the standard procedure for the optimization study.

Under an argon atmosphere, a solution of BuLi in hexane (1.6 mol/L 0.40 mmol, 0.25 mL) was slowly added to a solution of dry

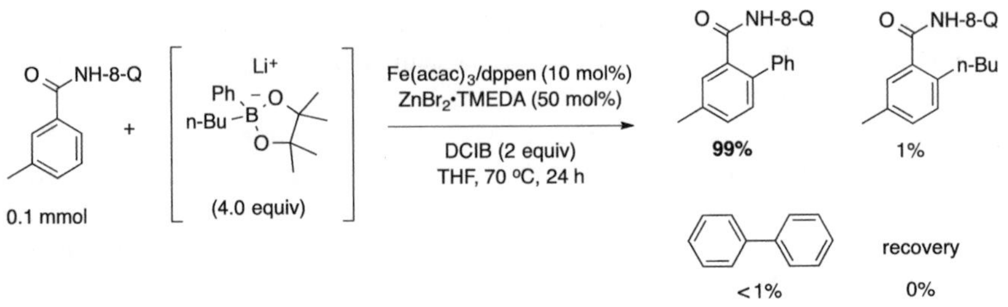

Scheme 11.1 Standard procedure for the optimization study

4,4,5,5-tetramethyl-2-phenyl-1,3,2-dioxaborolane (PhBPin, 0.40 mmol, 82 mg) in THF (0.5 mL) using a gastight syringe at −78 °C. The resulted mixture was stirred at −78 °C for 10 min and then allowed to warm to room temperature, where it was stirred for an additional 20 min to form the lithium borate salt. Next, a solution of Fe(acac)$_3$ (0.01 mmol, 3.5 mg), (Z)-1,2-bis(diphenylphosphino)ethene (dppen, 0.01 mmol, 4 mg), ZnBr$_2$ · TMEDA (0.05 mmol, 17 mg) and 3-methyl-N-(quinolin-8-yl)benzamide (0.10 mmol, 26.2 mg) in THF (0.5 mL) was transferred to the borate solution via cannula or gastight syringe at r.t. under argon. The resulted orange mixture was stirred at r.t. for 10 min, then 1,2-dichloroisobutane (0.2 mmol, 23 μL) was added. The mixture was stirred at 70 °C for 24 h (after heating for 20–30 min, the color of the mixture became black). After cooling to r.t., tridecane (0.1 mmol) was added as an internal standard. Aqueous potassium sodium tartrate was then added, and the organic layer was extracted with diethyl ether and passed over a Florisil column. The mixture was analyzed by GC using tridecane as an internal standard, or analyzed by ^{1}H-NMR using 1,1,2,2-tetrachloroethane as internal standard.

Effect of Moisture and Air Contamination
For the reaction on a small scale (0.1–0.2 mmol scale), moisture contamination can completely stop this reaction, presumably by poisoning the iron catalyst. For reproducible results, it is recommended that the reagents and glassware used to be thoroughly dried, and the reaction performed under a rigorous argon atmosphere.

   NOTE: Usually, the reaction color turns black after heating at 70 °C for 10–20 min. When moisture contaminated the reaction mixture, the color of the reaction mixture failed to turn black upon heating (remained orange, see the picture below), and in that case no desired product was formed.

Investigation of the Additive
The reactions were performed according to the General Procedure using various additives (Scheme 11.2). After aqueous work up, the crude mixture was analyzed by ^{1}H-NMR in the presence of 1,1,2,2-tetrachloroethane as an internal standard.

Investigation of the Amount of Zinc Additive
The reactions were performed according to the General Procedure using different amounts of ZnBr$_2$ · TMEDA (Scheme 11.3). After aqueous work up, the crude

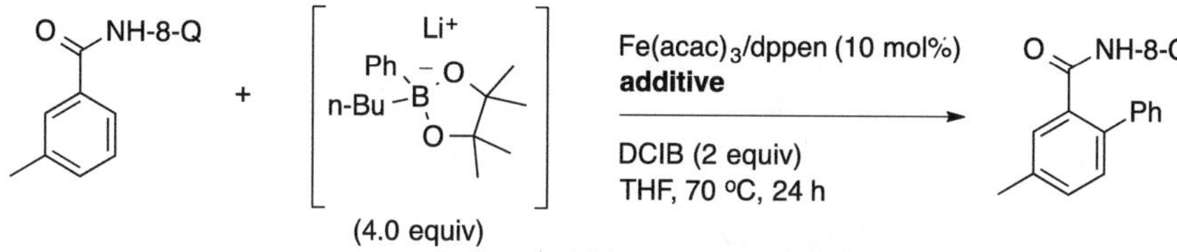

Scheme 11.2 Investigation of the additive

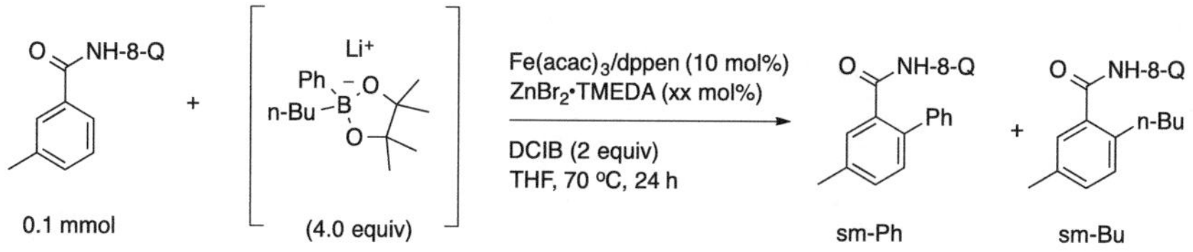

additive = **ZnBr$_2$·TMEDA (20 mol%)**   **96%**
ZnBr$_2$·TMEDA (50 mol%)   99%
ZnBr$_2$·TMEDA (100 mol%)   99%
ZnCl$_2$·TMEDA (50 mol%)   99%
ZnBr$_2$ (50 mol%)   95%
Zn(OTf)$_2$·TMEDA (100 mol%)   0%
MgBr$_2$·Et$_2$O (300 mol %)   0%
none   0%

	sm-Ph	sm-Bu	recovery	Ph-Ph
ZnBr$_2$·TMEDA (20 mol%)	96%	<5%	0%	<5%
ZnBr$_2$·TMEDA (30 mol%)	98%	<5%	0%	<5%
ZnBr$_2$·TMEDA (40 mol%)	99%	<1%	0%	0%
ZnBr$_2$·TMEDA (50 mol%)	99%	<1%	0%	0%
ZnBr$_2$·TMEDA (100 mol%)	99%	<1%	0%	0%

for sm-Bu and Ph-Ph, yields are GC yields based on calibration

Scheme 11.3 Investigation of the amount of zinc additive

mixture was analyzed by 1H-NMR in the presence of 1,1,2,2-tetrachloroethane as an internal standard.

Investigation of the Iron Precatalyst
The reactions were performed according to the General Procedure using several iron salts (Scheme 11.4). After aqueous work up, the crude mixture was analyzed by ^{1}H-NMR in the presence of 1,1,2,2-tetrachloroethane as an internal standard.

Investigation of the Ligand
The reactions were performed according to the General Procedure using various ligands (Scheme 11.5). After aqueous work up, the crude mixture was analyzed by ^{1}H-NMR in the presence of 1,1,2,2-tetrachloroethane as an internal standard.

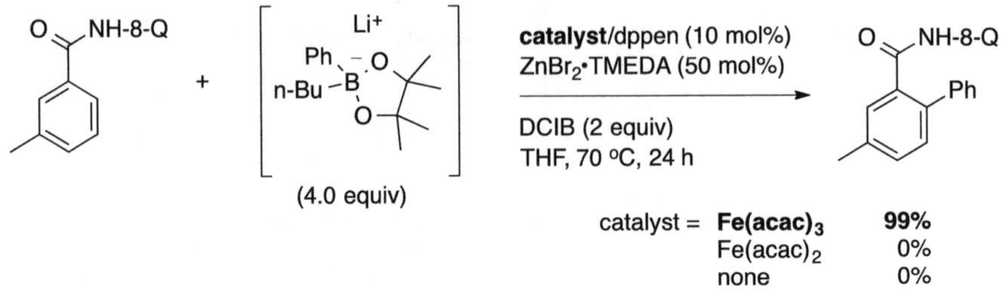

Scheme 11.4 Investigation of the iron precatalyst

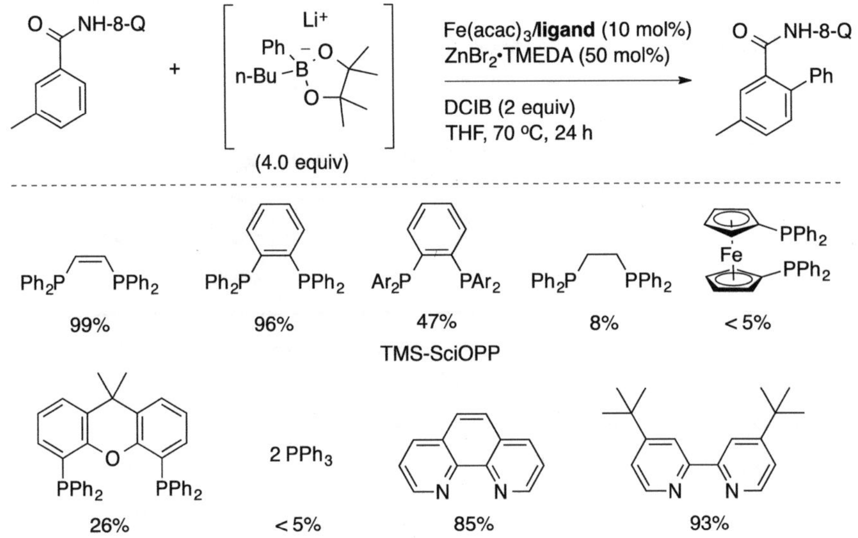

Scheme 11.5 Investigation of the ligand

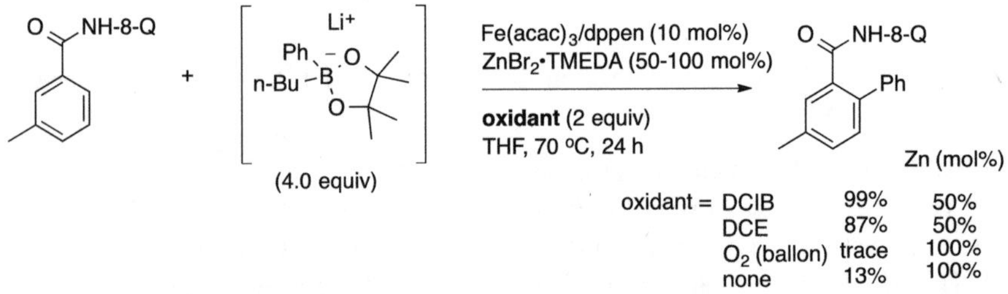

Scheme 11.6 Investigation of different oxidants

Investigation of Different Oxidants
The reactions were performed according to the General Procedure using different oxidants (Scheme 11.6). After aqueous work up, the crude mixture was analyzed by ¹H-NMR in the presence of 1,1,2,2-tetrachloroethane as an internal standard.

Investigation of Temperature
The reactions were performed according to the General Procedure at different temperatures (Scheme 11.7). After aqueous work up, the crude mixture was analyzed by ¹H-NMR in the presence of 1,1,2,2-tetrachloroethane as an internal standard.

Investigation of the Amount of Boronate
The reactions were performed according to the General Procedure using different amounts of boronate (Scheme 11.8). After aqueous work up, the crude mixture was analyzed by ¹H-NMR in the presence of 1,1,2,2-tetrachloroethane as an internal standard.

Investigation of the Effect of Organometallic Base for Activating the Boronate
The reactions were performed according to the General Procedure using different bases to activate PhBPin (Scheme 11.9). After aqueous work up, the crude mixture was analyzed by ¹H-NMR in the presence of 1,1,2,2-tetrachloroethane as an internal standard.

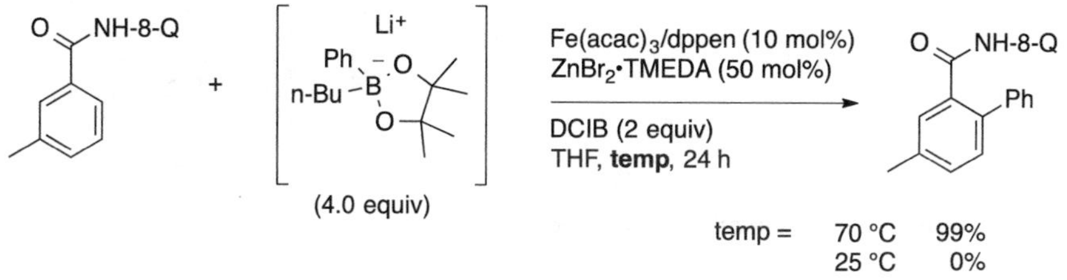

Scheme 11.7 Investigation of temperature

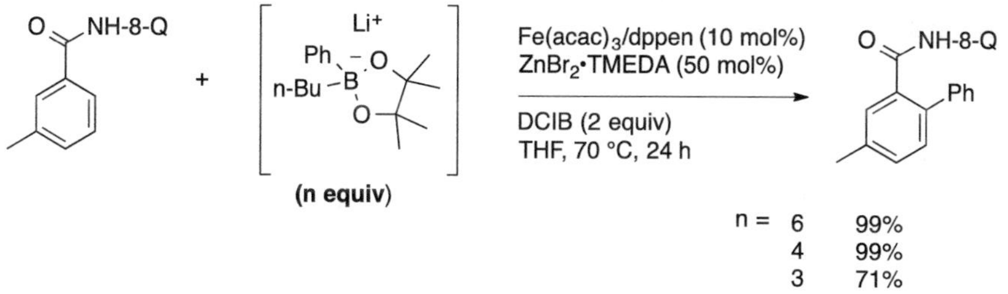

Scheme 11.8 Investigation of the amount of boronate

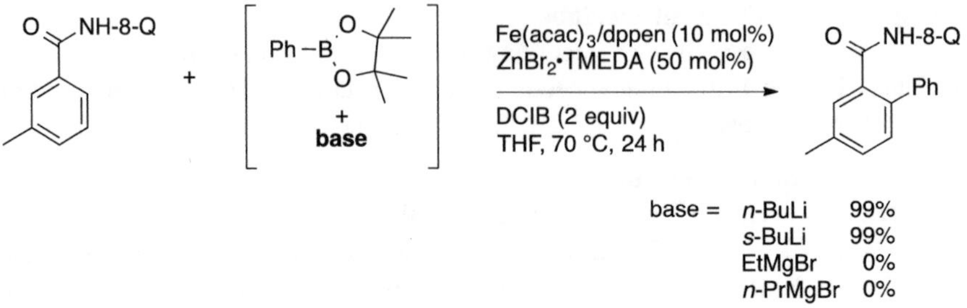

Scheme 11.9 Investigation of the effect of organometallic base for activating the boronate

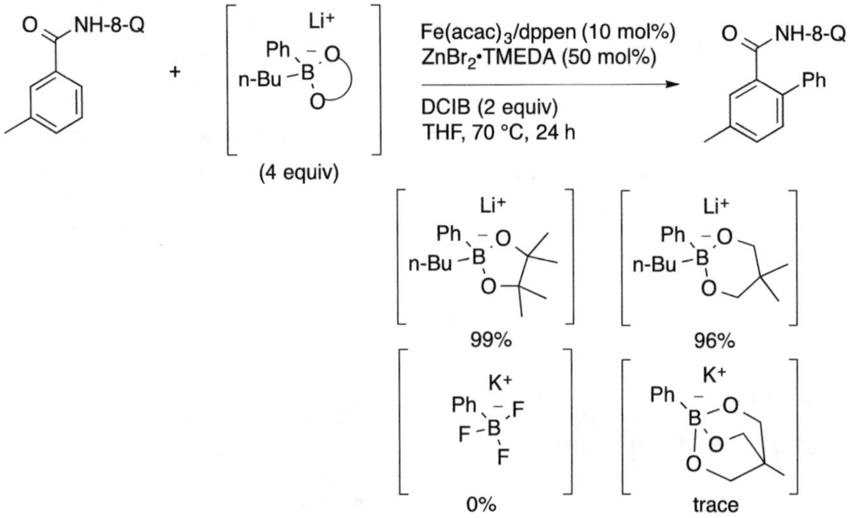

Scheme 11.10 Investigation of different borate reagents

Investigation of Different Borate Reagents

The reactions were performed according to the General Procedure using different borate sources (Scheme 11.10). After aqueous work up, the crude mixture was analyzed by ^{1}H-NMR in the presence of 1,1,2,2-tetrachloroethane as an internal standard.

11.4.2.2 General Procedure for Iron-Catalyzed Cross-Coupling of C(sp^2)–H Substrate with Organoboronate

Under an argon atmosphere, a solution of BuLi in hexane (1.6 mol/L, 1.6 mmol, 1.0 mL) was slowly added to a solution of dried R–BPin (1.6 mmol) in THF

(1.0 mL) using a gastight syringe at −78 °C. The resulted mixture was stirred at −78 °C for 10 min and stirred at r.t. for 20 min to form the lithium borate salt, then a red solution of Fe(acac)$_3$ (0.04 mmol, 14 mg), (Z)-1,2-bis(diphenylphosphino) ethene (dppen, 0.04 mmol, 16 mg), ZnBr$_2$ · TMEDA (0.4 mmol, 136 mg) and corresponding quinolinyl carboxamide (0.4 mmol) in anhydrous THF (1.0 mL) was transferred to the borate solution via cannula or gastight syringe at r.t. under argon. The resulted orange mixture was stirred at r.t. for 10 min, then 1,2-dichloroisobutane (0.8 mmol, 92 μL) was added via syringe. The mixture was stirred at 70 °C for 16–24 h (always after heating for 20–30 min, the color of the mixture becomes black). After cooling to r.t., aqueous Rochelle's salt was added (1.0 mL), and the organic layer was extracted with diethyl ether (5 mL × 3). The combined organic layers were washed with brine, passed through a pad of Florisil, dried over MgSO$_4$, and concentrated under reduced pressure. The crude mixture was purified by column chromatography on silica gel (hexane:ethyl acetate = 15:1– 30:1) to afford the corresponding product.

11.4.2.3 General Procedure for the Stoichiometric Experiments

Under an argon atmosphere, a solution of BuLi in hexane (1.6 mol/L, 0.80 mmol, 0.50 mL) was slowly added to a solution of dried 4,4,5,5-tetramethyl-2-phenyl-1,3,2-dioxaborolane (PhBPin, 0.80 mmol, 164 mg) in THF (1.0 mL) using a gastight syringe at −78 °C. The resulted mixture was stirred at −78 °C for 10 min and stirred at r.t. for 20 min to form the lithium borate salt, then a red solution of

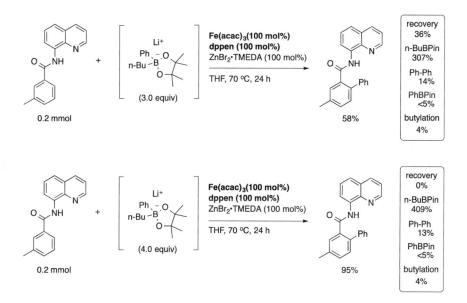

Scheme 11.11 Reaction using stoichiometric amount of Fe(III) without oxidant

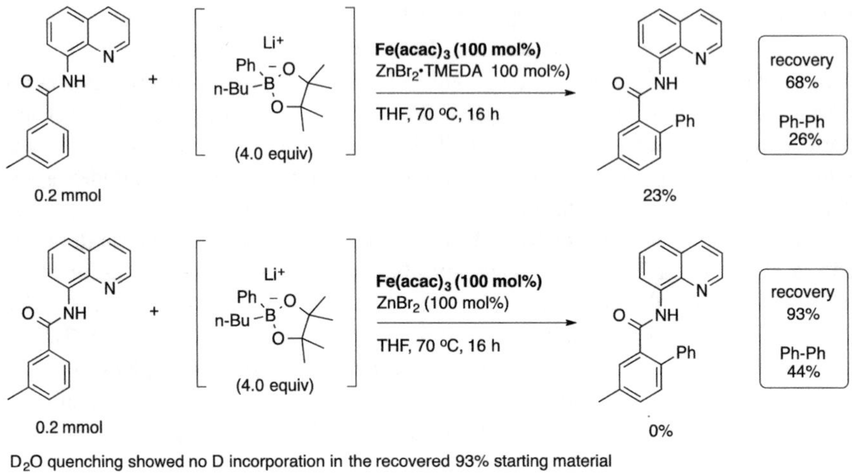

recovery 68%

Ph-Ph 26%

0.2 mmol   23%

Fe(acac)$_3$ (100 mol%)
ZnBr$_2$·TMEDA 100 mol%)

THF, 70 °C, 16 h

(4.0 equiv)

Fe(acac)$_3$ (100 mol%)
ZnBr$_2$ (100 mol%)

THF, 70 °C, 16 h

recovery 93%

Ph-Ph 44%

0.2 mmol   0%

D$_2$O quenching showed no D incorporation in the recovered 93% starting material

Scheme 11.12 Stoichiometric experiment in the absence of ligand (TMEDA and dppen)

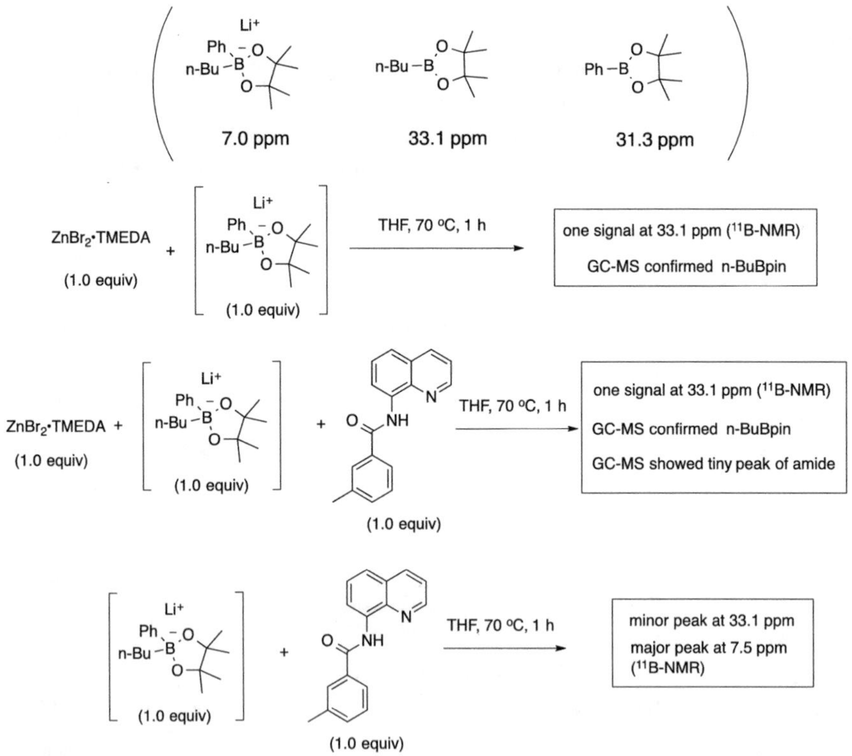

7.0 ppm   33.1 ppm   31.3 ppm

ZnBr$_2$·TMEDA +
(1.0 equiv)

(1.0 equiv)

THF, 70 °C, 1 h

one signal at 33.1 ppm (^{11}B-NMR)

GC-MS confirmed n-BuBpin

ZnBr$_2$·TMEDA +
(1.0 equiv)

(1.0 equiv)

(1.0 equiv)

THF, 70 °C, 1 h

one signal at 33.1 ppm (^{11}B-NMR)

GC-MS confirmed n-BuBpin

GC-MS showed tiny peak of amide

(1.0 equiv)

(1.0 equiv)

THF, 70 °C, 1 h

minor peak at 33.1 ppm

major peak at 7.5 ppm
(^{11}B-NMR)

Scheme 11.13 ^{11}B-NMR study

Fe(acac)$_3$ (0.20 mmol, 71 mg), (Z)-1,2-bis(diphenylphosphino)ethene (dppen, 0.20 mmol, 80 mg), ZnBr$_2$ · TMEDA (0.20 mmol, 68 mg) and 3-methyl-N-(quinolin-8-yl)benzamide (0.2 mmol, 52.4 mg) in anhydrous THF (1.0 mL) was transferred to the borate solution via cannula or gastight syringe at r.t. under argon. The resulted black mixture was stirred at r.t. for 10 min and then it was stirred at 70 °C for 24 h. After cooling to r.t., tridecane (0.2 mmol, 49 µL) was added as an internal standard. Aqueous Rochelle's salt was then added, and the organic layer was extracted with diethyl ether and passed over a Florisil column. The mixture was analyzed by GC using tridecane as an internal standard or analyzed by ^{1}H-NMR using 1,1,2,2-tetrachloroethane as internal standard (Schemes 11.11 and 11.12).

11.4.2.4 ^{11}B-NMR Study

^{11}B NMR spectra were recorded at 96.3 MHz. Chemical shifts for boron are reported in parts per million (ppm) and are referenced to the external standard boron signal of BF$_3$ · OEt$_2$ (THF-$d8$, = 0.00 ppm) (Scheme 11.13).

References

1. *Handbook of C–H Transformations: Applications in Organic Synthesis*, Dyker, G., Eds. (Wiley-VCH, Weinheim, 2005).
2. Sun, C.-L., Li, B.-J., & Shi, Z.-J. (2010). *Chemical Communications. 46*, 677–685, and references therein.
3. Nishikata, T., Abela, A. R., Huang, S., & Lipshutz, B. H. (2010). *European Journal of Organic Chemistry, 132*, 4978–497.
4. Wasa, M., Chan, K. S. L., & Yu, J.-Q. (2011). *Chemistry Letters, 40*, 1004–1006.
5. Engle, K. M., Thuy-Boun, P. S., Dang, M., & Yu, J.-Q. (2011). *Journal of the American Chemical Society, 133*, 18183–18193.
6. Tredwell, M. J., Gulias, M., Gaunt Bremeyer, N., Johansson, C. C. C., & Collins, B. S. L. *et al.*, (2011). *Angewandte Chemie International Edition, 50*, 1076–1079.
7. Thuy-Boun, P. S., Villa, G., Dang, D., Richardson, P., & Su, S. *et al.* (2013). *Journal of the American Chemical Society, 135*, 17508–17513.
8. Ueno, S., Chatani, N., & Kakiuchi, F. (2007). *Journal of Organic Chemistry, 72*, 3600–3602.
9. Chinnagolla, R. K., & Jeganmohan, M. (2012). *Organic Letters, 14*, 5246–5249.
10. Vogler, T., & Studer, A. (2008). *Organic Letters, 10*, 129–131.
11. Karthikeyan, J., Haridharan, R., & Cheng, C.-H. (2012). *Angewandte Chemie International Edition, 51*, 12343–12347.
12. Waterman, R. (2013). *Organometallics, 32*, 7249–7263.
13. *Iron Catalysis in Organic Chemistry*, Plietker, B., Eds. (Wiley-VCH, Weinheim, 2008).
14. Nakamura, E., & Yoshikai, N. (2010). *Journal of Organic Chemistry, 75*, 6061–6067.
15. Sun, C.-L., Li, B.-J., & Shi, Z.-J. (2011). *Chemical Reviews, 111*, 1293–1314.
16. *The Chemistry of Organoiron Compounds*, Marek, I., Rappoport, Z., Eds. (John Wiley & Sons, Ltd., Chichester, UK, 2014).
17. Nakamura, E., Hatakeyama, T., Ito, S., Ishizuka, K., & Ilies, L., *et al.*, *Organic Reactions*, S. E. Denmark, Ed. (John Wiley & Sons, 2014), vol. 83.

18. Nakamura, E., & Sato, K. (2011). *Nature Material, 10*, 158–161.
19. Tolman, C. A., Ittel, S. D., English, A. D., & Jesson, J. P. (1978). *Journal of the American Chemical Society, 100*, 4080–4089.
20. Klein, H.-F., Camadanli, S., Beck, R., Leukel, D., & Flörke, U. (2005). *Angewandte Chemie International Edition, 44*, 975–977.
21. Camadanli, S., Beck, R., Flörke, U., & Klein, H.-F. (2009). *Organometallics, 28*, 2300–2310.
22. Bedford, R. B., Hall, M. A., Hodges, G. R., Huwe, M., & Wilkinson, M. C. (2009). *Chemistry Communications*, 6430–6432.
23. Kobayashi, Y., & Mizojiri, R. (1996). *Tetrahedron Letters, 37*, 8531–8534.
24. Norinder, J., Matsumoto, A., Yoshikai, N., & Nakamura, E. (2008). *Journal of the American Chemical Society, 130*, 5858–5859.
25. Yoshikai, N., Matsumoto, A., Norinder, J., & Nakamura, E. (2009). *Angewandte Chemie International Edition, 48*, 2925–2928.
26. Ilies, L., Asako, S., & Nakamura, E. (2011). *Journal of the American Chemical Society, 133*, 7672–7675.
27. Asako, S., Ilies, L., & Nakamura, E. (2013). *Journal of the American Chemical Society, 135*, 17755–17757.
28. Matsubara, T., Asako, S., Ilies, L., & Nakamura, E. (2014). *Journal of the American Chemical Society, 136*, 646–649.
29. Valio, I. F. M., Burdon, R. S., & Schwabe, W. W. (1969). *Nature, 223*, 1176–1178.
30. A Mg(II) salt under similar conditions did not effect the transmetalation: Hatakeyama, T., Hashimoto, T., Kondo, Y., Fujiwara, Y., & Seike, H. *et al.*, (2010). *Journal of the American Chemical Society, 132*, 10674–10676.
31. Bedford, R. B., Gower, N. J., Haddow, M. F., Harvey, J. N., & Nunn, J. *et al.*, (2012). *Angewandte Chemie International Edition, 51*, 5435–5438.
32. Zaitsev, V. G., Shabashov, D., & Daugulis, O. (2005). *Journal of the American Chemical Society, 127*, 13154–13155.
33. Rouquet, G., & Chatani, N. (2013). *Angewandte Chemie International Edition, 52*, 11726–11743.
34. Shang, R., Ilies, L., Matsumoto, A., & Nakamura, E. (2013). *Journal of the American Chemical Society, 135*, 6030–6032.
35. Webster, C. E., Fan, Y., Hall, M. B., Kunz, D., & Hartwig, J. F. (2003). *Journal of the American Chemical Society, 125*, 858–859.
36. Vastine, B. A., & Hall, M. B. (2007). *Journal of the American Chemical Society, 129*, 12068–12069.
37. "The Nobel Prize in Chemistry 2010", *Nobelprize.org*.
38. Blanchard, S., Derat, E., Desage-El Murr, M., Fensterbank, L., & Malacria, M. *et al.*, *Eur. J.* (2012). *Inorganic Chemistry*, 376–389.

Printed in the United States
By Bookmasters